THEMISTIUS
On Aristotle's "Physics 4"

THEMISTIUS
*On Aristotle's
"Physics 4"*

Translated by
Robert B. Todd

Cornell University Press

Ithaca, New York

Preface and Appendix © 2003 by Richard Sorabji
© 2003 by Robert B. Todd

All rights reserved. Except for brief
quotations in a review, this book, or parts
thereof, must not be reproduced in any form
without permission in writing from the publisher.
For information address Cornell University Press,
Sage House, 512 East State Street, Ithaca, New York 14850.

First published 2003 by Cornell University Press.

ISBN 0-8014-4103-X

Acknowledgments

The present translations have been made possible by generous and imaginative funding from the following sources: the National Endowment for the Humanities, Division of Research Programs, an independent federal agency of the USA; the Leverhulme Trust; the British Academy; the Jowett Copyright Trustees; the Royal Society (UK); Centro Internazionale A. Beltrame di Storia dello Spazio e del Tempo (Padua); Mario Mignucci; Liverpool University; the Leventis Foundation; the Arts and Humanities Research Board of the British Academy; the Esmée Fairbairn Charitable Trust; the Henry Brown Trust; Mr and Mrs N. Egon; The Netherlands Organisation for Scientific Research (NWO/GW). The editor wishes to thank David Furley, Stuart Leggatt and Pamela Huby for their comments, and John Bowin for preparing the volume for press.

Printed in Great Britain

Librarians: Library of Congress Cataloging-in-
Publication Data are available

Contents

Preface *Richard Sorabji*	vii
Introduction	1
Place, Void and Containers	7
Textual Emendations	9
Translation	15
Place: Chapters 1-5	17
Void: Chapters 6-9	36
Time: Chapters 10-14	52
Notes	75
Bibliography and Abbreviations	112
Appendix: The Commentators *Richard Sorabji*	116
English-Greek Glossary	127
Greek-English Index	130
Index of Passages Cited	145
Subject Index	148

To the memory of
Henry J. Blumenthal

Preface

Richard Sorabji

Henry Blumenthal, to whose memory this volume is dedicated, asked, and answered in the affirmative, a very important question in an article on Themistius which he entitled, 'Themistius: the last Peripatetic commentator on Aristotle?' Themistius was an orator, politician and essayist of the fourth century AD, in the capital city of Constantinople, and devoted only some of his time to philosophy in a privately run institution, so was not obliged to take sides between schools. But in the introduction to my forthcoming Sourcebook, *The Philosophy of the Commentators, 200-600 AD*, I shall answer Blumenthal's question by saying that he does side with contemporary Neoplatonism too much to be considered a member of the Peripatetic, i.e. Aristotelian school, e.g. requiring Platonic Forms to explain biological reproduction (so Devin Henry, see the translator's bibliography), and rejecting Aristotle's empiricist account of concept formation (*in DA* 3,31-4,11).

The fact that Themistius' commentaries are called paraphrases should not be allowed to conceal their importance. Thomas Aquinas famously appealed to another commentary by Themistius in his controversy with the followers of Averroes who denied individual immortality to the human intellect. Aristotle could be enlisted in the service of the Christian belief in individual survival, in the light of Themistius' commentary on Aristotle's *On the Soul*. More than that, as Robert Todd says in his introduction to the present translation, Themistius offers information and ideas not found elsewhere. In fact, some of his ideas are very original. He is also taken very seriously by later commentators. I shall illustrate this from the present commentary, starting with that on *Physics* Book 3, and progressing to that on Book 4.

Themistius, commenting on the preceding Book 3 of Aristotle's *Physics*, solves a problem that confused other commentators. Aristotle in this book defines change. He had described a change (*kinêsis*), such as walking a mile, as incomplete until the mile is done, yet at *Physics* 3.1, 201a10-11, he calls change a completion (*entelekheia*) of what is potential. How can it be both incomplete and a completion?

Themistius saw, 69,7-20, that it matters crucially what is meant by 'what is potential'. The completion of the bronze matter of a statue as something having the capacity to *have* been sculpted, would be a statue. But the completion of the bronze as something having the capacity to be *in process of being* sculpted would be a change (*kinêsis*), namely, the process of being turned into a statue.

In *Physics* Book 4, Aristotle defines place and time, and denies the possibility of vacuum. Themistius here has two tussles with Galen, the great doctor-philosopher of the late second century AD. First, we hear at 149,4-19 that Galen regarded time as self-revealing and accused Aristotle of circularity in his attempt to define it. Aristotle had defined it as involving the possibility of counting changes and marking them as before or after. Galen complains that this has to be understood as the before and after of *time*, the very thing that was supposed to be being defined. In response, Themistius first offers Aristotle's own defence, that the appeal is to the before and after of *position*, not of time. This is not entirely satisfactory, because positions are only thought of as before or after in relation to an imagined movement which reaches one position *chronologically* before another. But Themistius adds his own reply, that if there is a circularity, it is benign, because it is only right and proper that the definition should mean the same as what is being defined. The great sixth-century Neoplatonic commentator Simplicius demurs in his *in Phys.*718,13-719,18.

At 114,7-12, it appears that Galen had replied to Aristotle's denial that there is such a thing as three-dimensional space, distinct from the three-dimensional volume of a body. Aristotle thought it was enough to describe a thing's place, roughly speaking, as its surroundings, the surroundings into which it fits exactly. To postulate space as well would give us too many three-dimensional entities. Themistius defends Aristotle at 133,31-135,1. It would be very odd, he says, if a vacuum, or even space, could exist where a body is, and penetrate right through the body. Extensions exclude each other, and that is why bodies exclude each other – because they are extensions.

Galen, however, argued for the reality of space distinct from a body's volume, by imagining a bronze jar whose contents shrink, without any other matter coming in to fill the gap. Themistius accuses Galen of begging the question, by assuming the very spatial extension, or gap, that he wants to prove. But in the early sixth century AD Philoponus was to defend Galen, *in Phys.* 576,12-577,1 (Corollaries; see p. 82 nn. 128-9 below), by saying that he only hypothesises that no matter comes in after the shrinkage.

At 163,1-7, Themistius criticizes Aristotle, rightly I believe, for saying that there would be no time if there were no soul to count off the different positions in a movement. Aristotle's definition of time as the countable aspect of change in respect of before and after should require only that change is *capable* of being counted, not that there is an *opportunity* of counting it, such as would be supplied by the existence of souls to do the counting.

Finally, Themistius offers ingenious ideas on how Aristotle's outermost sphere which carries the stars can, on Aristotle's view, have any place. For Aristotle, a thing's place is its surroundings, and the outermost sphere has no surroundings. Themistius suggests among other things, 121,1-4, the solution that the outermost sphere could have as its place the surface of the next sphere in, that of the planet Cronus or Saturn. Simplicius (*in Phys.* 590,27-32; 592,25-7; see p. 86 n. 188 below) and Philoponus (*in Phys.* 565,21-566,7) reject this, partly on the ground that such a place would not be of equal size. Philoponus adds that the surface of the next sphere in cannot provide a place to the outermost sphere, when it also receives its place from there.

Introduction

The paraphrases of Themistius (*c*. 317 – *c*. 388 AD) belong both to the history of philosophical exegesis and to the history of philosophical pedagogy.¹ They were designed to clarify the texts of some central works of Aristotle, and thereby make them accessible to relatively advanced students. The voice of the paraphrasist is often that of Aristotle, sometimes identical, but normally operating in a more expansive, yet sometimes more summary, mode, occasionally altering the expository order of a text, and, in rare cases, speaking independently through digressions, queries, or excursuses that clarify texts considered poorly organised or obscure.²

These paraphrases, then, were not introductory works (*eisagôgai*), but targeted at readers who wished to revisit Aristotelian treatises with which they were already familiar, and were pitched at a level somewhere between earlier expansive commentaries (notably those of Alexander of Aphrodisias, fl. *c*. AD 200) and strictly elementary expositions. Problems are introduced for expository purposes rather than criticism;³ earlier authorities and other Aristotelian works are rarely cited; and there are almost never major historical or critical digressions. Nonetheless they provide information and interpretation often not found elsewhere, and Themistius was sufficiently respected by later commentators to be widely quoted, discussed and translated. To the modern reader of Aristotle a text-by-text exposition may seem alien to the practice of critical synthesis that marks contemporary scholarship in the history of ancient philosophy. Even so, paraphrase represents a permanently useful *entrée* to difficult texts; it is as crucial a part of interpretation, today as it was in the first half of the fourth century AD.

Themistius' treatment of Aristotle's *Physics* concentrates on Books 1-4, the more accessible part of the treatise. He approximately doubles the length of the original text, whereas for Books 5-8 the paraphrase is *in toto* about 25% shorter.⁴ The version of Book 4 translated here, like the rest of the *Physics* paraphrase, is less adventurous than Themistius' major philosophical work, the paraphrase of Aristotle's *de Anima*. Clearly, the topics of place, void and time did not represent for him an opportunity for interaction with Neoplatonism, or for critical analysis, such as we find in the exegeses of *Physics* 4 from the early sixth century AD by Simplicius and Philoponus.[5] His treatment is probably fairly close to the orthodoxy maintained in Alexander of Aphrodisias' lost *Physics* commentary,[6] where Aristotle's discussions were also undoubtedly linked with Hellenistic philosophy in greater detail.[7]

The present translation (the first of any part of this paraphrase since the Latin version of the whole work by the Venetian Ermolao Barbaro the Younger [1454-93], published in 1481)[8] is based on the edition in *CAG* by the Austrian scholar, Heinrich Schenkl (1859-1919), best known for his preliminary work on the Teubner edition of Themistius' orations. He canvassed a wide range of manuscripts, and wrote an elaborate *Praefatio* (seemingly one of the most elaborate in the whole *CAG*). He fixed on four manuscripts, each of which served to represent a sub-group, and one of which had, in his opinion, the greatest authority.[9] He also used the Aristotelian text and its variant readings, and drew on the traces of the Themistian paraphrase, both explicit and implicit, in the commentaries on the *Physics* by Simplicius and Philoponus. There were also two earlier printed editions: the Aldine (1534),[10] and one by Leonhard Spengel (1803-80) published in 1866.

While I have not consulted any manuscripts, I have had to change the selection from variant readings in Schenkl's text, and to introduce some emendations. I have also had to readjust punctuation, which was too often uncritically inherited from Spengel's edition.[11] Unlike Themistius' commentary on the *de Anima* where an Arabic translation preserved a valuable independent tradition,[12] the text of that on the *Physics* is derived solely from Greek manuscript sources. I have tried to improve them by using the Aristotelian text and in particular by exploiting material from Philoponus and Simplicius who represent the best indirect tradition for the Themistian text.

In the translation a major problem was presented by the frequently used cognate noun and verb *kinêsis* and *kineisthai*.[13] They sometimes refer to motion in the sense of locomotion, but more often to the generic process of change (also identified by *metabolê/metaballein*, which I have done as 'transformation'/'be transformed').[14] Locomotion (for which the dedicated noun and verb are *phora* and *pheresthai*) is a species of change, one aspect of 'change in respect of place' (increase and decrease being the others; cf. 111,20-1; 123,25). My policy has been to translate *kinêsis* as 'movement' where it refers to locomotion, and *phora* only as 'motion'. When *kineisthai* refers to locomotion (as notably in the paraphrase of *Physics* 4.8), it is translated as 'move'. The verb *pheresthai* can in most cases be translated 'be carried' (particularly given its use to describe the passive motion of elements or inanimate bodies),[15] except for one section of Chapter 11 (150,15-151,24; cf. n. 448) where 'be in motion' is more appropriate in the context of a general argument about time. All cases in which 'movement' and 'move' are used for *kinêsis/kineisthai* are separately recorded in the Greek-English Index.

Secondly, *khronos*, 'time', presents a special problem since it covers both the general concept of time, and, often with qualifying terms, a specific time-period. Since there is no separate Greek term for a time-period, my policy has been to use this expression only within the restricted context of the arguments about motion in the void in *Physics* 4.8, and to exclude it in the major discussion of time (4.10-14) where the text is best left transparent rather than be subjected to any interpretive translation. Certainly the term 'time-period' is not found in modern translations of Aristotle.

Abbreviated references in the notes to secondary literature, and to collections of primary sources, are explained in the Bibliography. Abbreviations used for the works of Aristotle and the commentators follow the practice of this series.[16] In the translation square brackets are used for my own clarificatory supplements, angle brackets (as in orthodox editorial practice) for emendations to the Greek text in the form of supplements; any of the latter that are unexplained can be assumed to be part of Schenkl's text. Round brackets are used simply for punctuation, though to a greater degree than would be normally acceptable in English prose, due to the extensive use of parentheses in this exposition; it is such a device to which the instruction 'bracket' refers in the list of Textual Emendations below.

Acknowledgments

In preparing this translation I have had the benefit of valuable criticism from David Furley, Stuart Leggatt, and particularly Pamela Huby, whose insights on the problems of time were notably helpful. Access to unpublished material was kindly granted by Devin Henry and Bob Sharples. Chris Morrissey's assistance was indispensable in preparing the Greek-English index, and John Bowin's and Inna Kupreeva's editorial contribution invaluable in the final stages. I also benefited from a research grant from my university.

I dedicate this work to the memory of Henry J. Blumenthal (1936-98), whose general assessment of Themistius remains fundamental.[17] Henry's patient and careful scholarship on the Greek Aristotelian commentators was exemplary, and he was always ready with help and encouragement. His work and his memory will endure.

Notes

1. On Themistius in general see Todd (6), 1-2, and Todd (8). On his paraphrastic method see in addition to Todd (6), 2-7 the case-studies by Cacciatore and Ciollaro, and see Pignani for an overview. The crucial evidence for his methodology is his *in An. Post.* 1,2-12, translated and discussed at Todd (6), 3, and his *Or.* 23, 89,20-90,5 (see Todd [6], 2-3), now available in a translation of the whole oration at Penella, 108-27 at 121-2.

2. For the two major digressions on place and void in the present work see *in Phys.* 113,30-116,12 (cf. n. 125) and 132,3-133,15. The classification of Themistian responses to the Aristotelian text can be endlessly elaborated, particularly if account is taken of the details of glossing or replacement of Aristotelian terminology. By correlating the Themistian paraphrase with relatively short Aristotelian texts in this translation, I hope to allow the reader to determine easily the type of paraphrase involved in any given case.

3. See, for example, 120,21-8; 149,26-150,10 (cf. n. 442); and especially 161,29-163,7 (cf. n. 544).

4. The ratio by word count of Themistius to Aristotle in the eight books is: 2.18, 1.84, 2.27, 1.99, 1.10, 1.12, 0.38, and 0.88. I am indebted to Chris Morrissey for the calculation. Ballériaux, 201 has a rough calculation by pages of the ratio of Themistian to Aristotelian content for the paraphrases of the *Posterior Analytics* and *de Anima*. The precise figures (again supplied by Morrissey) are: 0.99 and 1.06 for *An. Post.* 1 and 2, and 2.63, 2.21 and 2.91 for *DA* 1-3, but 27.18 for *DA* 3.5, the chapter on the intellect.

5. Simplicius' commentary on *Physics* 4 is translated by Urmson (1) (4.1-5 and 10-14), (2) (Corollaries on place and time), and (3) (4.6-9). For Philoponus' Corollaries on place and void see Furley. Simplicius quotes Themistius at length only for the problems that he constructed in *Phys.* 4.14; see below nn. 545, 547 and 551.

6. On this commentary see Sharples (2), 1185, and Moraux (3), 129-80, with 621 for literature on the remains in Arabic.

7. While Themistius mentions and criticises the Epicureans and Stoics, he perhaps surprisingly fails to cite the Peripatetic Strato of Lampsacus (d. *c.* 268 BC), with whom he was familiar from Alexander's commentaries (see his *in de Caelo* 50,33-51; Wehrli (2), fr. 53, and Gottschalk [1] nos. 3a and 3b for Stratonian material from the same commentary). Strato's ideas about a micro-void could well have been introduced into the paraphrase of chs 6-9, as they were by Simplicius (*in Phys.* 652,19-25 and 693,11-18).

8. On Themistius' fortuna in the Renaissance and Barbaro's translations in particular see Todd (8).

9. These are **B** (Breslau [Wroclaw], Magd. 1442; s. xiv), **L** (Parisinus graecus 1886; s. xvi), **M** (Modena, Biblioteca Estense *a*.M.9.13 [II.A.4]; s. xiv), and **W** (Venice, Biblioteca Marciana 205; s. xv), the readings of which are reported in detail; see Schenkl, Praef. xxxviii. The editor did not construct a stemma, but argued (Praef. xxxv-vi) that MS **M** had the best authority, a position open to challenge (see, for example, the Textual Emendations and notes below *ad* 105,1; 107,1; 108,4; 148,5; and 152,24-5). (On the Greek manuscripts of Themistius' paraphrases in general see Todd [7]).

10. The copy of the Aldine was one at Munich that contained notes by Petro Vettori (1499-1585); Schenkl (Praef. xxxvii) inspected this for himself. On the Themistius Aldine see Sicherl, 8-10.

11. I have recorded changes where I felt that the printed text was notably

insensitive to sentence structure. I have, of course, on occasion silently changed punctuation for the purposes of translation.

12. See Todd (6), 10; for subsequent work on this Arabic translation see Browne (1) and (2).

13. For similar reflections see Urmson (1), 9 n. 1.

14. I have translated *alloiôsis* as 'alteration' (its usual equivalent in recent translations) to avoid any confusion with other terms for change. On the distinction between *kinêsis* and *metabolê* (which Aristotle uses interchangeably in *Physics* 4; see 218b19-20) see *Phys.* 5.1, 225a34-b3.

15. Its single use to describe the motion of the planetary spheres at 119,22 (*kuklôi pheresthai* = *periphersthai*, 120,18) should not cause problems.

16. See Sorabji (2), 12-17 for full details.

17. See Blumenthal (1) and (2).

Place, Void and Containers

In his discussions of place and void Themistius, like Aristotle, makes frequent references to containers. The abstract term for container (*periekhon*) is often matched by generic terms for vessels (*angeion*; *skeuos*), and occasionally by references to three specific artefacts.

(1) The amphora is a pot (or *kaddos*) used to carry wine, and so a helpful illustration of an enclosed space the contents of which can change, and which can itself be repositioned (116,12-118,1).

(2) The bulbous *kratêr* serves to illustrate the kind of container that cannot fit flush against a wall, and so might be erroneously thought to show that there is bound to be a void space between bodies (114,21-4).

(3) The shallow *kylix* is the perfect container for the ashes that proponents of the void claim can absorb water into their interstitial void spaces (127,21-2).

The *kylix* would also be the best container from which to demonstrate natural evaporation, or mutual replacement, of water by air, from which we derive a basic notion of place (102,21-103,2), though any boiling pot could do the job.

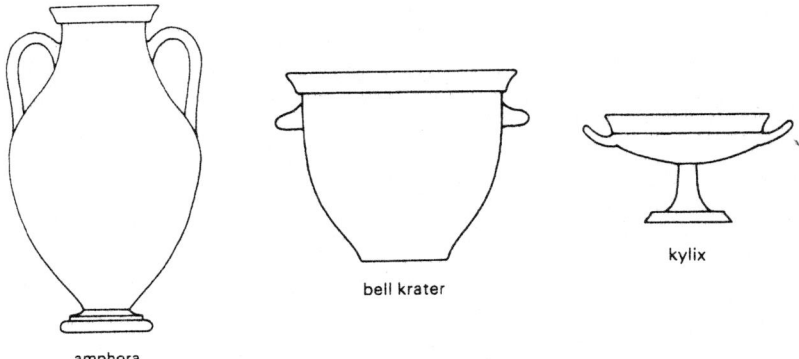

amphora   bell krater   kylix

Textual Emendations

102,15	For *kath'hauto* read *kath'hauta* (Vat. gr. 1025; Spengel)
103,13	For *metastanti* read *metastasi* (Spengel)
103,26	Comma after *eige estin*
104,8	For second *to* read *tôi* (MS **W**; Spengel)
104,12	Close direct speech after *periakhthêsêi*; delete quotation mark at 104,17 after *sômati*
104,13	Question mark for comma after *topon*; change *ti* to an interrogative form
104,25	Correct *hosa* to *hora*
104,33	For *periekhon* read *periekhôn* (Philop. *in Phys.* 511,25)
105,1	For *isôs de on* read *isos de ôn* (MS **W**; Philop. *in Phys.* 511,25)
105,13	Correct *hê* to *ho*
105,22	For *legomen* read *legômen*
107,1	For *einai ton toutôn topon* read *toutôn einai ton topon* (MS **W**; Arist. 209b22)
107,10	Stop for question mark after *epiphaneia*
108,1	Colon after second *topon*; remove brackets from *phtheiretai – apodounai*; stop after *metabolais* (108,2)
108,4	For *autou* read *autês* (MSS **WBL**; Spengel)
108,6	For *legomen* read *legômen* (MS **W**)
108,10	Supply *ho* before *anthrôpos* (Spengel)
109,15	Comma for stop after *leukon*; for *hôs* read *hôste* (Furley)
109,25-6	Delete *kai hai phuseis kekhôrismenai*
110,3	For *enginetai* read *eginêtai* (MS **W**)
110,16	Delete *autôi houtô* and *ho amphoreus*
111,2	Supply *hôs* before *en topôi*
112,2	Stop for comma after *sunekhê*
112,6	For *kath'auto* read *kath'auta* (MSS **BL**)
112,15	Delete *hôs*
112,15	For *tou eskhatou* read *tôi eskhatôi* (corrector of MS **W**)
112,21	Delete colon after *topoi*; bracket *meizô – sômatôn* followed by comma
112,22	Bracket *hê – keramiou*
113,7	Stop for comma after *arithmos*
113,29	For *ekrhuseie* read *ekrheuseie* (Usener)

Textual Emendations

113,30	For *labêi* read *laboi* (Simpl. *in Phys.* 573,19)
114,5	For *to dê* read *tôi de*
114,12	For *oute* read *ouden*
114,14	For *to legein* read *tou legein*
114,14	Commas around *ei tis hupothoito*
115,26	Colon for comma after *autôn*
115,28	Supply *en* before *tôi tou topou* (MS **L**)
116,11	Supply *ti* before *asaphesteron*
116,13	For *en heautôi* read *en tautôi* (Philop. *in Phys.* 550,10); comma after *menein*
116,14	Delete *ê aeros*
116,20	For *diastêmata* read *diastêmatos* (MS **W**; Philop. *in Phys.* 550,17)
116,26	Delete *tou hudatos*
119,8	Supply *to* before *para to meson*
119,20	Delete stop after *kineitai*; bracket *hautê – topos* (21); comma for stop after *topos*
121,5	Delete colon after *dunamei*; bracket *pôs – holês* and follow with comma
121,15	Supply *to* before *pan* (Arist. 212b18)
122,10	For *to* read *tôi* (coni. Schenkl)
122,27	Delete *touto*
123,25	Bracket *kinêsis – phthisis* (cf. Ross *ad* 213b5)
123,32	Insert question mark after *allo*
124,2	Stop for comma after *ginetai*
124,24	For *legomen* read *legômen* (Vettori)
125,17	Supply *ti* after *tode* (Arist. 214a12)
126,10	Delete colon before *apo*; bracket *apo – sômatôn* (10-11) followed by a comma
126,14	For *kai tauton* read <*kai*> *tauton*
126,17	For *exetazomen* read *exetazômen* (Spengel)
126,18	Delete colon after *legousin* (18); bracket *ou gar – elathen* (19) followed by comma
126, 22	Delete colon after *kinêseôs* (20); bracket *dunatai – kekhôrismenou* followed by stop
126,29	For *ekpurinêzontos* read *ekpurinêzomenou* (MS **M**)
127,13	For *hora de* read *hora dê* (Vettori)
128,5	Supply *ou* before *mallon*
128,23	For *an* read *au*
129,1	For *antikeimenêi* read *antikeimenê*
129,8-9	Delete colon before *to – Dêmokritos* and bracket it
129,11	Add comma after *tautês*
129,15	Comma for stop after *phusin*
129,27	Comma for stop after *kenôi*
130,3	Stop for question mark after *pou*
130,4	Delete *hôs te*

Textual Emendations 11

130,6	Question mark for comma after *iskhuroteron*
130,6	Colon for stop after *logos*
131,1	For *ison* read *auto*
131,14	Comma for stop after *khronos*; comma after *kineitai* moved to follow *diastêma*
132,17-18	For *kai* read *kei* (Shorey, 448; *ei* MS **W**); delete the question mark in 132,18
133,1	Stop for comma after *kinêthêsetai*
133,1	For *tên hupothesin* read *têi hupothesei*
133,8	For *allôs* read *oxeôs*
133,10	For *tauta* read *tauta* (with crasis) (Shorey, 448, who cites 133,11)
133,24	For *aulêtôn* read *aulôn*
133,29	For *ekbibasaio* read *embibasaio* (Shorey, 448)
134,30	Delete *to* before *sôma*
135,6	For *kakeinou allo* read *kakeino allou* (see 4.8, n. 326)
135,23	Replace the colon preceding and stop following *hoson – hepetai* with brackets
136,1	Delete *kenon*
136,1	For *eroito* read *heloito* (MS **W**)
136,1	For *kumainei* read *kumaneî* (sc. future tense)
137,27	Delete *epi*; ignore lacuna
138,11	For *autê hê hulê* read *hê autê hulê* (Arist. 217a28)
138,17	Comma for question mark after *stoikheiois*
138,24	Delete *to* before *dunamei*
138,25	Delete *einai*
139,1	For *all' hoti* read *all' allo ti* (Simpl. *in Phys.* 689,10)
139,15	For *to* before *di' holou* read *tôi*
139,30	Bracket *oute – manois*
140,13	For *kai mên kai eniautos* read *kai hêmera mên eniautos* (Simpl. *in Phys.* 698,10-11)
140,14	Transpose *oude – estin* (140,15) to follow *khronos*; bracket it; begin a new sentence at *to*
140,14	Delete *anankaiôs*
141,15	For *mellei* read *melloi* (MS **W**)
141,28	Comma for question mark after *hen*
141,29	Question mark for colon after *eniautos*
141,31	Supply *hen* before *peras* (Arist. 218a23)
142,7	For *dêlon hoti* read *dêlon ti*
142,25	Delete *autou*
143,9	Delete *ou*
143,27	Question mark for comma after *exeuroi*
143,27	Comma after *ennoêsômen*
143,28	Restore *pote*
144,1	Delete *pote*
144,13	Stop after *khronou*

Textual Emendations

144,25	Delete colon after *khronon*; bracket *houtô – Aristotelên*
146,1	For *apo* read *epi* (MS Med. Laur. 85,14)
146,5-6	Delete colon after *sunuparkhei* (145,5) and comma after *todi* (146,6); bracket *hama gar todi to sêmeion kai todi*
146,7	For first *proteron* read *husteron*
146,7	Replace comma after *ên* with stop
146,8	Delete colon and *hôsper ge* after *ekeino*
146,10	Comma for colon after *tropon*
147,8	Lacuna after *alla tina thesin*; see 4.11, n. 424
147,14	For *dis* read *tis*
147,21	Supply *to men* before *proteron* (corrector in MS **L**)
148,5	For *houtô legein* read *houtô legein ekhei* (MSS **WSL**)
149,23	Comma after *khronoi*; reposition question mark after *pantakhou*
150,6	Delete question mark after *nun*
150,25	For *auto gar to* read *to gar* (Simpl. *in Phys.* 723,33)
150,28	Delete stop after *nun*; bracket *en huparxei – tode*
150,29	Supply *ti* after *tode* (cf. 150,27)
151,19	Supply *lambanomenon* after *sêmeion*
151,27	For *to nun* read *tôi nun*
152,4	Question mark for stop after *peras*
152,16	Delete *hoion*
152,19	Stop for question mark after *grammôn*
152,21	Delete colon after *to elakhiston*; bracket *elakhiston – toiauta* (151,22), placing a colon after *grammai*, and a comma after the bracket
152,23	Delete colon after *hexei*; bracket *hekaston – diaireton*
152,24-5	Delete *hai duo hôrai*; comma for colon after *estin*; bracket *ou – touto* (25)
152,26	Delete *de* after *touto*
153,5	Supply <*hôi men oun arithmoumen*> (Torstrick) for the lacuna
153,9	For *palin* read *palin kai palin* (MS **W**, Arist. 220b13-14; Simpl. *in Phys.* 733,1)
153,22	Delete *kai ton hena anthrôpon*
154,9	Begin brackets at *ou* not *ê*
154,15	Supply *tôn* before *autôn*
154,21	Replace colons around *hôrismenos gar ho horismos* with brackets; follow with comma
154,23	Delete colon after first *khronos*; bracket *kai – tis*
156,1	For *epeidê* read *epei de* (*epeide* [sic] MS **M**)
156,22	Delete colon after *ou*; bracket *ou gar – husteron* (23)
157,7	Stop for comma after *khronôi*
157,27	Colon for comma after *diaphora*
158,10	Question mark after *pote*
158,11	Stop for question mark after *khronos*

Textual Emendations

158,14	Supply *an* before *eiê* (MSS **MWB**; Arist. 222b5-6)
159,8	Supply *ouk* before *arkei* (P.M. Huby)
159,14	Delete *tên* before *metabolên* (MSS **BL**)
159,16	Supply *en* before *genesei* (MS **W**; Simpl. *in Phys.* 755,2).
160, 5	For *toutôn* read *tauta* (MS **W**); comma after *exôthen*; stop deleted after *êkribologeisthô*
160,7	Delete comma after *husteron*; delete *estin*
160,8	Delete colon after *apostasin*; bracket *to – mellontos*; comma for colon after *mellontos*
160,28	Transfer closing quotation mark from *arithmountos* (line 27) to follow *aisthanomenou* (line 28)
160,29	Supply *to* before *arithmêtikon*
161,22	Delete *hepta* and the comma following it
161,22	Supply *epi* before *tôn kinêseôn*
161,24	For *kata tauta* eliminate the crasis of *ta auta*
161,24	For *arithmoumenoi* read *arithmoumenai*
162,19	For *hôs* read *hôste* (Simpl. *in Phys.* 766,6)
163,1	For *homoiôs* read *homoiou* (Shorey, 448)
163,5	Question mark for stop after *endidontos*
163,19	Delete first *khronos*

Themistius
On Aristotle Physics 4

Translation

Themistius' Paraphrase of Book Four of Aristotle's *Physics*

[Place: Chapters 1-5]
[Chapter 1]

102,2 (208a26-9) Place, like [the infinite],[1] must also be investigated by the natural philosopher in the same sequence: first, as to whether it exists; then as to the way in which it exists (whether as a substance, i.e. *per se*, or as having its being in something else); also, as to what in general terms it is.[2]

102,6 (208a29-32) Traditional belief[3] also shows that the discussion of place belongs to this kind of inquiry, since it assumes that everything that exists is somewhere. But an invalid reasoning formulates this assumption. For since *anything that does not exist is nowhere* (a goat-stag and a sphinx are, after all, nowhere), people also think, through ignorance of the logical conversion, that it follows that *anything that does exist is somewhere*. But even if they err in the [latter] universal posit,[4] it is certainly true that the natural bodies (earth, fire, water, air, plus what consists of them) are in place, so that this natural reality must also be discussed by the natural philosopher. And, to generalise, if he can inquire into change, and if the primary and most widespread change for all natural bodies is the one in respect of place,[5] then by that token he could not also neglect the investigation [of place].

102,14 (208a32-208b1) Now its definition is difficult, since nobody inquiring into what [properties] *per se*[6] belong to place could derive an identical definition from all of them, as will be clear quite soon. But we even lack a partner to inquire into our presuppositions, since nobody has found a way of reaching the truth about place by solving its problems; in fact, nobody has even adequately formulated them. But still, using whatever means possible, we must not shirk the task.

102,20 (208b1-8) Now the belief that place in general exists could come from bodies replacing one another: i.e. where previously there was water, air is now present, because the water has flowed out, as it does in vessels,[7] **[103]** and so the same place remains, but receives successively different bodies. From this two things can be confidently asserted: the existence of place, and its being different from what it receives.[8]

103,4 (208b8-22) Also, each [natural] body is carried into its proper place (fire up, earth down) as long as it is not impeded. So this demonstrates not only that place is something, but that it also has certain distinguishing marks of its own, and, in effect, powers[9] – *viz.* up and down, and the remaining directions towards which bodies move. Nobody, that is, should believe that up and down, backwards and forwards, and left and right exist only in their relation to us, but should realise well before this that they are also distinguished in nature, and the difference is obvious. What is up and down, that is, are often identical in their relation to us (e.g. the ceiling of the house is currently above us, but when we ascend to the roof it is below us, and the pillar is currently on the right, but after we change position,[10] is on the left). On the other hand, what is *naturally* up and what is *naturally* down are separated by the movements of the natural bodies, up being what fire (i.e. what is light) moves towards, down being what bodies that are heavy (i.e. contain earth) move towards. And Nature not only sets these [locations] well apart in terms of their position relative to one another, but also in terms of their power (for they have the capacity to receive things that are differentiated by vast distances).

103,18 (208b22-5) What becomes particularly clear is that up and down relative to us and as distinguished by nature are not identical, unlike the objects of mathematical study, which *per se* have no place and are not moved anywhere (for they also do not exist *per se*), but relative to us can be conceived and drawn as left and right, but not so as[11] to have each of these [directions] by nature, but through a conception derived from us.[12] That is, the way that we conceive of them is the way that we also imagine place along with them, although they have no place by nature.

103,25 (208b25-7) Also, those who introduce the void introduce it as place; that is, we conceive of the void, if it exists,[13] as simply place stripped of body.

103,26 (208b27-209a2) These [arguments] suffice to prove that place is something, yet the tribe of poets also should not be discredited, particularly Hesiod who even depicted place as prior to everything: the famous *khaos*, that is, is intended to signify for him that there must be a prior space for what exists.[14] (In fact this man, it seems, is one of those who think that **[104]** everything that exists is somewhere.)[15] If Hesiod's claim is correct, then the nature of place emerges as something extraordinary: i.e. as prior to everything that exists, granted that it is a prior condition[16] for everything. Also, while nothing else exists without place, *khaos* exists even without anything else, because it preexisted other things in time. Yet place is also prior by its nature, since it joins in eliminating [occupants], but is not jointly eliminated with them; i.e. is not destroyed when its contents are. For example, if fire were destroyed, its place is not also destroyed, but if place is destroyed, its contents[17] are also destroyed along with it, assuming that being nowhere and not existing are identical.[18]

104,9 [*ad* 209a2-4] [19] But see that we do not inflate[20] place beyond its due. For consider the counter-arguments too, which, as well as adding nothing to place, actually eliminate it once and for all. 'Try,' they say, 'to define what place is, and you will be brought round to agreeing that it does not exist.'[21]

[six arguments against the existence of place][22]

104,12 (209a4-7) (i) First, to what genus will you assign place?[23] To body obviously, in that place is extended in three [directions].[24] But in that way the greatest absurdity[25] of all will crop up: a body will go through a body totally, and two bodies will occupy the same place. For if place and the body that enters it are bodies, and both equal in their dimensions, then one body will be in another body of equal [volume]. Chrysippus and Zeno's followers[26] have this [consequence] <in>[27] their doctrines, though earlier thinkers reduced it to a virtually self-evident impossibility, and many have engaged with this argument at the level of the genus [body], including Alexander of Aphrodisias in his treatise *On Mixture*, and in his commentaries on natural philosophy, as well as some others who can be easily encountered.[28]

104,22 (209a7-14) But (ii) if place is not a body, it could scarcely emerge as something different, given that being extended in three [directions] is exclusive to body. But place receives water from air; this, after all, is what we believe proves the existence of place.[29] But recognise[30] that this [process] too does not avoid some absurdities that are hard to handle. For when the place in which there was previously air receives water, clearly it also receives the surface of the water, as previously it did that of the air. But if it receives the surface, it will also receive the lines, and receive the points too. So there will be a place even for a point, just as there is for a body. But that is impossible. For why will the place of the point differ from the point?[31] And [the place of] the line and of the surface will not [differ] either– for what will the place of the surface be apart from the surface itself? For it is not larger than it[32] (for any part by which it extends beyond it contains[33] none of it, though place depends **[105]** on this [containment]), but through being equal [in size][34] it is simply[35] a surface. But if it is a place for the line, it will not differ from the line, and on receiving a surface will not differ from the surface, and by containing a point, is itself also a point, and, when it becomes the space for a body, it will become a body.

105,4 (209a14-18) (iii) In general terms, each thing that exists is either an element or consists of elements, whether it is said to be a body or an incorporeal. But while the elements of bodies are also bodies, and the elements of incorporeals are incorporeals, place is included in neither: not among the intelligibles, since it has magnitude; not among bodies, since, as we have said, it receives bodies.

105,9 (209a18-23) But (iv) to say that place is also prior and like a first principle is beyond accidental error. For how will it be a first principle? As matter is? So what is compounded from it? But is it like form, or like what causes change? Or like that for the sake of which? Yet in what way?[36]

105,12 (209a23-5) (v) Zeno's problem too could be rationalised:[37] that if place is included in what exists, it would also itself be in a place,[38] as would the[39] place of the latter place, and so on *ad infinitum*.

105,14 (209a26-9) But (vi) if the void were completely eliminated (to correspond with the truth),[40] then if place exists, clearly it would never be devoid of body; instead, every place is completely filled. But if that is true, what are we going to say about things that increase? From what source, that is, will a body that becomes larger acquire place? For all [bodies] are already in occupancy.[41] Now the place necessarily gets larger when it is increased in volume along with the body. So how, and from what source, will a place be increased? It is difficult to give an explanation.

105,20 (209a29-30) So it *seems* to have been correctly claimed that anyone trying to define what place is will have to agree that it does not exist.[42]

[Chapter 2]

105,22 (209a31-3) So once more let us make a new start and state[43] that one kind of place is spoken of as primary,[44] another in respect of something else: in respect of something else is the one common to multiple bodies; primary the one proprietary to each [body] in isolation.[45]

105,24 (209a33-b1) I could say that the whole house as well as the *polis* is your place in respect of something else, indeed the Earth too, or even the air and the cosmos.[46] But what is your *primary* place? It is the one that contains nothing more than you. In fact, the sequence will be truly ascribed because of it: i.e. if you are in the house, it is because you are in *this* place in it, which coincides with you, and if you are in the *polis*,[47] it is because you are in the house, and if you are on the Earth, it is because you are in the *polis*.

105,29 (209b1-2) But if **[106]** place in its strict sense is what primarily contains each body, then place will be a specific limit: for it is at this that bodies are delimited, and at it primarily (certainly not at something else prior to it), as well as at it proximately[48] (for there is nothing in between).[49]

106,4 (209b2-5)[50] But if place is also a limit, both primary and proximate, then the form of each [body] emerges as its place too, for in a primary sense it delimits and bounds each single body, including the matter underlying physical magnitudes. So from this perspective the form might seem to be place.

106,7 (209b5-11) But if we conceived of some other property belonging to place, matter would clearly be more associated with place. That is because a place is held to be the magnitude's (i.e. the body's) extension, and it is as something like this that matter too is envisaged, for when bounded by form and contained by planes and limits, it becomes a magnitude (i.e. a body). *Per se* it is an extension, but an unbounded extension: for when the limit and the incidental properties are abstracted from a sphere, nothing is left beyond the matter and the extension. That is because the extension is not in the limit, but is bounded by [the matter and the limit], yet is in the substrate rather than in them. This, then, is why place has an affinity to matter, namely, because while remaining the same it receives multiple structures, i.e. the forms that are in bodies.

106,18 (209b11-17)[51] On the basis of this similarity Plato actually says in the *Timaeus* that matter and space are identical: i.e. he speaks of the participant in forms (namely matter) and of space (namely place) as identical. Yet in the *Timaeus* and in the unwritten doctrines he says that matter receives the forms in different ways: by participation (*methexis*) in the *Timaeus*, by assimilation (*homoiôsis*) in the unwritten works.[52] But still, as I said, he would seem to be claiming that matter and place are identical. For while everyone was saying that place was *something*, he alone tried to explain *what* it was. (But he seems to have used the term metaphorically, for he offers no further clarification.)[53]

106,26 (209b17-21) However, the difficulty of the investigation is also clear from the fact that whether place is form, or indeed matter, its definition is problematical, since it reverts to first principles, into which insight is in itself very abstract and not straightforward, since we are trying by an act of separation to get to know things [sc. form and matter] that cannot be known separately from one another. **[107]**

107,1 (209b21-8) Yet it is not difficult to see that neither [form nor matter] can be place,[54] since, unlike place, they are inseparable from the physical object. For example, what air used to be in is identical with what water in turn enters, when, as we have said,[55] these bodies replace one another. For in general place is also not something belonging to a body – not a part, an affection, a *hexis*, or a power –[56] since none of these are separated [from a body], whereas place can be separated from each of them.

107,6 (209b28-33) In fact, place seems to be somewhat like a vessel, in that the vessel is a transportable place[57] yet the vessel does not in any way belong to the object. Thus it might be asked: isn't [place] an incidental property of the body, if it is also separable? Well, it is *not* incidentally related to the body from which it is separated, but is the limit and surface of the [body] that is contained.[58] And it could not be matter because of yet another argument: i.e. that [matter] is contained, whereas [place] contains.[59] And again, place is distinct from, and external to, the physical object,[60] whereas neither the matter nor the form are like that.

107,13 (209b33-210a2) As for Plato, if he really intends to make matter place, then observe that he says that both the forms and the eidetic numbers[61] are in a place.[62] In fact, as I said, he posits matter as underlying these, and this he sometimes names 'the great and the small',[63] at other times 'the capacity for participation'.[64]

107,16 (210a2-5) This aside, if bodies that move naturally are carried towards their proper places, but not carried towards their matter and form at all (each of these being with itself and in itself), neither of the latter pair will be place. But, if either is place, I would certainly be glad to know how in general bodies could also move, if they always have their proper place with them. Furthermore, how will up and down be distinguished in places? Or how will there be movement in these directions? What, after all, is up and down in matter? What are they in form? Yet an inquiry into place must address the directions towards which bodies have natural movement.

107,24 (210a5-9) But assume that either matter or form is going to be place, and that they move together with bodies and come to coexist where the physical object is – then, first, it is absurd for places to move; and next, how will a place not be in a place? That is, if a body enters a place when it undergoes a transition from here to there, clearly the form too is in a place; after all, it does change position along with it.

107,28 (210a9-11)[65] Now when bodies are transformed into one another, and water comes from air, then if you are going to say that the form is the place **[108]**, you destroy the place. That is because in transformations the form ceases to be.[66] But it is difficult to explain what the ceasing to be of a place is. But if [place] is the matter, why does water enter another place, although it is not displaced from its matter, i.e. not separated from it?[67] For it is this [matter], [Aristotle] says, that remains when it receives different forms.[68]

108,4 (210a11-13) This, then, can conclude our preliminary statement of problems.

[Chapter 3]

108,6 (210a14-15) Let us state[69] next the number of ways in which 'in something' is used, since that might offer some help to the discussion.

108,7 (210a15-24) In one way (i) it is like[70] the finger in the hand, and, generally, like the part in the whole; in another (ii) like the whole in the parts, as the face is in the eyes, nose and mouth; in another (iii) like the species in the genus (as <the>[71] human being is in Animal); and also (iv) like the genus in the species (as Animal is in the human being). (And don't be surprised if Animal is said to be in the human being, as the greater in the lesser; for with accurate understanding, you would find that Animal is in the human being essentially as a part – it is, after all, included in its definition.) 'In something' is also spoken of (v) as the form in the matter (the bench in the wood; health in what is hot and cold). (vi)[72] is as the [affairs] of the Greeks are 'in [the power of] the King', i.e. in general terms, in [the power of] the first cause of change (cf. Homer's 'it lies *upon* the knees of Zeus').[73] (vii) is as 'in the final purpose, i.e. the good' – [e.g.] 'everything for me lies *in* happiness'. There are still other senses of 'in something', but of those stated, as well those omitted,[74] the fundamental one is (viii) that according to which 'in something' is spoken of as in a vessel, i.e. as in a place.

108,21 (210a25-33) With these distinctions drawn, the next inquiry must be into whether something can be in something in the sense of in itself, and whether this sense falls under one of those that have been mentioned, or whether it is totally impossible for something by itself[75] to be in itself, with the truth instead being that everything is either nowhere, or in something else in the way that we say that the human being is in himself. Now it must be realised that we say that the soul of a human being is in the human being's body not primarily, but in respect of something else,[76] i.e. that the part of this whole is in another part. For only in this way can something by itself be in itself, **[109]** since the part is in the part (as the animal is also said to be self-moved because one part of it causes movement while another is moved), but otherwise a whole cannot be in a whole.

But we do not say (someone claims) that the soul is in the body but that the animal is in itself. This person seems unaware of the licence in ordinary usage, since it names many things just from the parts. So just as it states that Socrates has shoes on, not because his whole body also does, but because his feet do,[77] so it also states that the animal is in itself because one part of it is in another. For I could also say that the amphora of wine (meaning the vessel made of clay plus the wine) is in itself because the wine is in the vessel, but not that the vessel is separately in the vessel, or that the wine is in the wine.

109,10 (210a33-b8) Thus something is by itself in itself only in respect of something else, i.e. in respect of a part. But many things are said to be in something in respect of something else: e.g. pallor is said to be in the human being because it is in the body, and in the body because it is in the surface, but in the surface not in respect of something else but in a primary sense. Yet it is not because the pallor is in the surface in a primary sense that a surface and its pallor are identical, and something by itself is consequently[78] able to be in itself; for there are distinct definitions for surface and pallor, and distinct natures for knowledge and the soul. Yet knowledge is in the soul in a primary sense; in fact, it is in a human being because it is in the soul.

109,18 (210b8-18) But still, if you inquired at the individual level,[79] you would find that everything that is in something in a primary sense is distinct in nature from the preceding [cases], and, if this is the case, how will something be in itself in a primary sense? For then it will be distinct from itself! But if in respect of one of the senses enumerated above,[80] then it *is* in itself. For everything that is distinct is in something distinct [from it]; e.g. when we say 'the whole is itself in the parts',[81] we do not also say 'it is in itself'. That is because the whole and the parts differ, the [whole] being one and the same as itself, the [parts] multiple and distinct. But not even the account we have given concedes that the essence[82] is different both from what is in something in a primary sense, and from what it is in primarily.[83] So if something came to be in itself, it would receive different definitions – specifically, different things will coincide. Yet just as wine cannot be simultaneously wine and a clay amphora, or the clay amphora simultaneously a vessel and wine, so neither can something by itself be in itself. But, as stated,[84] perhaps something could all by itself[85] be in itself in respect of its parts, **[110]** as one might also speak of 'the amphora of wine' to allow one part (i.e. the clay amphora) to receive the wine, though it is not itself wine but distinct from the wine, and another (i.e. the wine) to come to be in[86] the vessel, not in the sense of in itself as a vessel, but [in the sense of] in a different thing. For in general the definitions of container and content[87] are totally distinct.

110,5 (210b18-22) So the preceding shows that something cannot be *per se* and in itself – but not even incidentally. What is incidental and what exists in respect of a part are not identical (even if both exist in respect of something else); the two are in fact quite distinct. An animal, for example, is said to see in respect of a part (i.e. the eyes by which it sees are a part of it); the mast is moved incidentally in the boat, since it is incidental to the boat that is in motion; and pallor changes incidentally because the human being whose property it is does. For there are the following two senses of 'incidental': either when the part of something is said to be affected in just the same way as the whole, or when the incidental property [is affected in the same way] as that to which it is incidental. So how could something be incidentally in itself? For then it would have to be either a part or an affection of itself, and that cannot even be conceived. And in that way something incidental would also be in itself, as the vessel[88] is also incidentally in itself,[89] if it is assumed to be a part or an affection of the water. For since the water is in the vessel, the vessel too will be in itself, as a part or affection of the water. But this is ridiculous. Aside from the assumption being absurd, two bodies will also be in the same place: the vessel, if it is in itself, and the water, if the water is also in the vessel. So enough of this.

110,22 (210b22-7) Zeno's problem, by which he thought that if there was place, it also had to be in something, can be solved from the senses of 'in something' given above.[90] That is because 'in something' has several senses, and so place will be in *something*, but not in *place*. For place, when used in its strict sense as the limit of the container,[91] is in something (indeed in the body of which it is a part or limit), yet not in the sense of in a place. Health, for example, is in something (namely, in what is hot and cold), but **[111]** as a *hexis*;[92] and heat and coldness are in something, but as an affection. By the same token, place is in something, but not <in the sense of being>[93] in place; this was the absurdity in the progression to infinity.[94]

111,3 (210b31)[95] This, then, will complete our statement of the problems.

[Chapter 4]

111,4 (210b32-4) Given that the nature of place is hard to pin down,[96] it might be clarified from our first investigating what the common notions[97] envisage for place *per se*.

111,6 (210b34-211a11) Now we demand of place (i) that it contain that of which it is the place; (ii) that it not in any way belong to the physical object that it contains (that has already been demonstrated);[98] (iii) that the primary place be neither smaller nor larger than its content[99] (it will not be smaller – for how will it be a container? But were it larger, it would receive still another body, so that it would not be primary, as well as not belonging to this [body] exclusively). In addition, [we demand] (iv) [that place] be left behind in separation from its content (that, after all, is how it was also initially[100] conceived on the basis of the mutual replacement of bodies); (v) that every place have an up and a down; and (vi) that each body be in one of these [positions].[101] With these [criteria] established, we must continue the investigation, and must supply just such an account of place as will solve the problems raised, safeguard the properties thought to belong to place, and display the reason for the contentiousness surrounding it.

111,17 (211a12-23) So first it must be realised that there would be no inquiry into place without change in respect of place. For example, we regard the heavens as being in place primarily because they more than anything else are involved in [such] change.[102] But there are three kinds of change in respect of place: motion, increase, and decrease. (In increase[103] there is also a change of one place into another, i.e. what was formerly there is in turn repositioned into a smaller or a larger [place].)

Next, not everything that changes advances us to the conception of place, only things that change *per se*; those that do so incidentally do not strictly speaking change, and so neither need a place, nor contribute to knowledge of it. Yet among things that change incidentally there is a difference between those that can also change *per se* (e.g. the parts of a body, such as the rivets on a ship, if they are discrete), and those that cannot change at all (e.g. pallor and knowledge). But still, neither the one nor the other leads us to the primary notions of place, but only things that change *per se*. **[112]** Changed *per se* are all things that while detached from their container are in contact with it, without being continuous with it.[104] The hand changes as it currently writes in separation from the rest of the body, yet it is not detached from the body, but continuous with it and not in contact with the shoulder. Therefore it also does not change *per se* in respect of this part, i.e. it is not in place. Instead, as I said, the nature of place has to be pinned down unambiguously from things that change *per se*.[105]

112,7 (211a23-b5) An animal is changed *per se* in the cosmos. So would you speak of the cosmos as the animal's place? Not at all. The air instead? Not even that. What, then, given that it is assumed that place must be neither larger nor smaller than the body?[106] So what is Callias' place? The extremity of the air that contains only Callias. This, after all, is why he is in the air, and in turn why he is in the cosmos, so that this is his primary place. So when the container is not discrete but continuous, he is not spoken of as being in place, but as a part in a whole (as I also described the hand in the body);[107] but when it is discrete and in contact [with him], he will be in a place that is the extremity of his container,[108] and this is neither a part of the body (it is, after all, on the outside), nor indeed greater or less, but equal everywhere to the limits of the body that it contains; for the extremities of bodies that are in contact coincide.[109] So the parts of what is continuous do not change in the whole but along with the whole, whereas [bodies] that are discrete are changed in place, not along with place – as long as they are not ones that are in vessels; for vessels are transportable. But they are not the primary places (after all, they are larger than the bodies)[110], but rather their extremities (the hollow surface[111] of the pot, i.e. that of the clay)[112] by which they coincide with[113] the water or the wine. If indeed our claims are true, they make it obvious what place is: namely, the limit (i.e. the extremity) of the container that is in contact with the containing body.

112,25 (211b5-10) This could be demonstrated in yet another way: for since place must be one member of a quartet – meaning form, matter, the extension between the extremities of the container, and the actual extremities of the container – then if it is none of the first three, the remaining one is necessarily place.

112,28 (211b10-14) That it is not matter has been adequately shown;[114] but that it is also not form either is by now clear from what duped those who believed place was form.[115] **[113]** For since both the formal structure and the place are equally containers, they believed that the form was the place through ignorance of the syllogistic reasoning. Look, form and place are both limits, yet not of the same thing, but form only of the thing that is contained, place of the containing body.[116] I would not even say that the form is a limit; for the form is not even a surface, but the definition of what it is to be, except perhaps when it delimits and bounds matter (it is said to be a 'limit' just as number is also [said to be] a measure).[117] So it remains for us to demonstrate that place is also not the extension.[118]

113,8 (211b14-19) An extension is what is conceived of as between the limits of the container, e.g. what is within[119] the hollow surface of the pot. Now this belief is traditional, and associated with those who posit the void, yet later both Chrysippus' crowd[120] and Epicurus[121] were nonetheless adherents. Some imposed the doctrine on Plato too.[122] It relies on a plausible explanation, yet one that is quite false: namely, since we reach a conception of place in general from the mutual replacement of bodies (i.e. from different bodies continually coming to be in the same place at different times), they took place to be the intervening extension, which they believed remained the same when it received the bodies that were replacing one another, while being separated from each of these incoming bodies.

Vessels above all egged them on to this inference. For since water and air enter the vessel at different times while the hollow surface within the clay remains the same (i.e. circumscribed by unique limits), they inferred the existence of the extension within the hollow surface, which resembled the surface of the vessel in remaining the same (i.e. separated from the bodies) as it received the bodies in succession. But this is invalid. If the vessel could at any time be devoid of body, then perhaps this so-called 'extension' would be detected *per se*. But, as it is, fluid flows out and air simultaneously enters to replace it, and that leads them astray. For since every body is accompanied by an extension, they transfer the extension belonging to bodies to place, without reasoning that an extension is always in place just because a body always is too, as completely covered bronze vessels reveal: for [in their case] there would be no efflux of fluid[123] unless the air acquired[124] a space for its influx.

[excursus: problems in defining place as extension]

113,30[125] What dupes them is that the vessels' hollow surface also always remains unconflated, so that were there an implosion[126] **[114]** when the fluid was extracted, as there is in the case of wine-skins, they would not be similarly deluded. Thus we do not similarly imagine that my body's extension stays behind in the air; for once it moves, the surface containing me is conflated (i.e. unified with itself), whereas in very solid vessels some other body is always in contact with the [surface], which is protected from conflation. And because[127] this body is in an extension they say that the surface too always has an intervening extension.

[*Galen and place as the void*]

114,7 But let us hypothesise that when the fluid was removed, no other body flowed in: a separate extension therefore remains within the surface. But the hypothesis is illogical, all-wise Galen,[128] for it hypothesises the very object of our inquiry. For while we are inquiring into whether it is possible for there to be a separate extension, you fabricate for yourself a picture of just what you want – that a separate extension exists – without proving that it exists.[129] In general terms, you conceive of something impossible:[130] for eliminating the mutual replacement of bodies is no different from completely eliminating body, nor indeed from saying,[131] on someone's hypothesis,[132] that neither the heavens nor the Earth nor any of the bodies currently existing will ever go on existing – that instead no body different from these even comes into existence.[133] In fact, this is the only way that [Galen] will get his wish to leave an extension in which there are bodies now, but not at another time. But this is impossible, and not what Galen intends: for an extension can never manage to subsist without a body – instead, the bronze that forms the hollow surface [of a vessel] would sooner implode than remain without a body.[134] (That the void[135] is really devoid of meaning will be demonstrated very shortly).

114,21 What then?[136] In the case of a wall surely you wouldn't believe that a body can never be fully in contact with it (unless you are admitting the void in this way),[137] but in the case of a *kratêr* [believe that it can] not?[138] Actually, something is always equally in contact with the [*kratêr*] too, and the difference is just in the shape. 'How then', he says, 'do we conceive of the extension as something different from the bodies?' But how do we do so for pallor and dark colour? It is by abstracting each of them from the body that we conceive of [the extension] *per se* – not as subsisting *per se* and separate from body, but just by separating it in definition. For since being water is not identical with water's having an extension of a given amount (the same applies to wine and air too), it is by abstracting through reasoning the qualities in respect of which **[115]** different things are water, wine and air that we conceive of the extension all by itself.[139] Then when it turns out that the three bodies enter the same container successively – bodies that differ in all their other qualities, but are equal to the same extension – we infer *a fortiori* from what we see that the extension is something different from the bodies, because the air that flowed in after the wine lacks the wine's other qualities, while retaining only the same extension as the wine. And since bodies are not in place in respect of their other qualities (e.g. heat or coldness) but only in respect of their extension, we duly believe that place *is* the extension – incorrectly, since it is not the case that if bodies are in place in respect of an extension, their place *is* the extension, or that since we move in respect of an impulse, movement *is* an impulse.[140]

The preceding [analysis] explains how those who have arrived at this doctrine were led astray.

115,13 As for the extension between the extremities of the container not being place, you need just recall everything that we determined as belonging to place when at the outset we adhered to the common notions:[141] *that place must contain*, whereas the extension must be contained; and *that place must be separated from the bodies*, whereas the extension will enter and leave along with each body. That is, we have said that someone demanding that place remain without bodies admits the void, even if unintentionally.[142] We also said that while the quantity of the extension also remains the same, it does not remain the same extension.[143] Further, we held that place did not have to belong to the body at all, but that the extension was virtually the body itself.[144]

115,22 In general, why does someone who says that the extension is place put the body in a place rather than the place in a body? For the extension is in the body more than the body is in the extension, given that one of them (being a quantity) is incidental, while the other (being a perceptible body) is a substance.

115,25 And I would be glad to learn this from them:[145] when a body enters a place, does it (a) enter the place by retaining its own extension (given that every body is in an extension), or (b) does it discard its own [extension] and only enter[146] that of the place? If they are going to say (b), then let them instruct us on how the extension ceases to exist when the body does not, and on how what is unique to [the body] ceases to exist, while something other than it accrues?

115,31 If they are going to say that *both* the extensions are preserved, that of the body and that of the place, how will the two coincide, and which of them will be in which? That is, which will you say is contained, and which contains? And how can two extensions coincide in the same dimension[147] in which just one of them used to be, but not two bodies? (The other incidental properties, after all, will be no impediment: for it is not heat, coldness, pallor and darkness that make the volume larger, **[116]** but only the extensions.) Now they admit that (*p*) [two dimensions coinciding] entails (*q*) [two bodies coinciding] – *we* can demonstrate that (*q*) is a falsity and jointly eliminate its antecedent (*p*).[148]

116,3 Furthermore, when a body increases, will the extension of the place also be increased, or will that of the body be increased rather than that of the place? Now, how could someone say that one of two magnitudes was increased when an addition was made to the other? For that is just like saying that a foal grows from the nutriment given a puppy. Yet it is rather difficult for an *equal* amount to remain. For how will there still be a place for the larger amount? This argument has no bearing on [place as] the limit of the container, for a surface *per se* admits no increase.[149]

116,10 That, then, is what *we* have to say on this doctrine; and while the [arguments] that Aristotle uses possess <some> degree of obscurity,[150] one must, as far as possible, try, as we proposed at the outset,[151] to uncover these too.

116,12 (211b19-25) If, he says,[152] there were an extension naturally capable of existing *per se* and permanently, then in the same place[153] there would be infinitely numerous places. Why ever so? Because when the vessel full of water[154] is transported elsewhere, the water's parts will act just like the whole of the water in the vessel. So just as the whole of the water that occupies its own extension moves together with the vessel that is being transported, so will each of the parts that occupies its own extension move together with the vessel. So when the amphora enters another place, it will obviously occupy an extension in between the extremities of the surrounding air – an extension of which the water will also be a part –[155] and the parts of the water will also occupy it.

But consider how many extensions enter the same [place]: (i) the one that the whole amphora occupies, (ii) the one that the water occupies as part of this, and (iii) the one that each of the parts of the water occupies; also (iv) the one that the whole of the water occupies within the amphora (for it is in the amphora as its place); and (v) the one between the extremities of the [amphora's] hollow surface; add (vi) the one that each of the parts of the water occupies. Add to these (vii) the one for the amphora, in that it belongs to a body [sc. the water],[156] and (viii) the one for the part in that it belongs to a body,[157] then you will discover multiple extensions in the same place.[158]

116,27 I despise as more absurd a single and distinct [extension] for the parts, since each of them will be in its own extension *and* in the [extension] of a larger part, and since division is to infinity, the addition of places will progress to infinity too. For in general *if* place is an extension, there is nothing absurd about each of its parts also being in place *per se*. In fact, each is *per se* in an extension, that extension being a part of the whole place. **[117]** In other words, for those who posit the limit of the container [as place] the part is in the place incidentally, since the whole is in place, but for those who [posit] the whole of the intervening extension [as place], surely the parts will be in place *per se*? Each one is therefore in an infinite number of places. That place also enters a place is obvious: for the extension between the hollows of the amphora will be in another extension once the amphora has changed position.

So do the same absurdities crop up for those who propose the extremities of the container? Not at all. Why? It is because I would not say that when the amphora is moved, water and its parts enter another *place*; for it is in the same surface, and will not exchange this, i.e. not acquire some other one as yet another extension.

117,11 (211b25-9) 'So how can the wine occupy the same place, and be moved from there to here (from the living-room into the storeroom) both as a whole and in respect of its parts?', someone will ask. It is because it is not also moved *per se* (for it does remain in the amphora); instead, the amphora in which there is wine is moved and changes position. For it is *that in which* there is air, water, or whatever, that changes position, not the air, water or their parts; instead, these are moved at some point if the [contents] in the earthenware interchange position with one another. But if the extension were a place, the bodies *per se* would change position along with the vessels, and their parts, by exchanging their extension just like them, would also enter one place after another *non-incidentally* – since each place is a part of the larger one, and the larger one in turn part of a larger one, and this right up to the largest place and extension, the one in which the cosmos is located. (In fact, what also happens to this astonishing theory is that it makes a place a part of a place, just as it does an extension of an extension. And that by way of digression is an additional consequence for them.[159])

But look: either they are either going to say that the wine *per se* is moved in the clay pots when they are conveyed elsewhere (and we have stated all the absurdities they will be admitting), or else that it is totally unmoved. And then what a joke for unmoved Thasian wine to be conveyed to Athens! For those who identify place with the extension will be unable to explain how, when the vessel is moved, the wine *per se* can remain unmoved, yet be moved incidentally.

This, then, is a sufficient argument **[118]** for place not being the extension.

118,1 (211b29-212a2)[160] That neither form nor matter [is place] has been demonstrated: for, our numerous arguments aside,[161] bodies must be at rest in their places (given that they are at rest in matter) and not separated [from them]. But the association of matter with place offers no similarity: in the case of [matter] we say '*what* was water before is now air'; in that of place '*where* water was before, there is now air'.

118,6 (212a2-7) But by now it is certainly agreed that if place is not one of the trio, matter, form and extension, then it must be the remaining member of the [original] quartet,[162] i.e. the limit of the container, at which it is in contact with the body that is contained.[163] And every body is in place, whereby it can change in respect of motion and alternatively be at rest. (But revolution and motion are not identical.)[164]

118,11 (212a7-14) Place is thought to be something immense and difficult to get knowledge of. That is because numerous intrusions cloud inquiry into it, and most of all the extension, given its considerable similarity, in that it too seems to remain unmoved, just like the primary place,[165] and to receive different bodies one after another. The cause of the confusion has been described:[166] it is that for the bodies [in place] the same thing seems to enter along with them and exit (due to their differing in other qualities, but being equal in extension, at least where they fill up an equal place),[167] although its substitution is disguised[168] because bodies replace one another rapidly.[169] Air's apparent incorporeality also contributes something by explaining why not only do the limits of the vessel seem to be place, but *ipso facto* everything in between too: for it is imagined to be empty because air does not provoke sensation like the other elements.

118,23 (212a14-21 + 28-30)[170] Place is not the whole vessel, only the extremities of its hollow surface, and it is also transportable, though not unqualifiedly so, just incidentally; for it is the body that is transported (and I mean the piece of clay known as the pot) and it has a limit. And just as the vessel is a transportable place, so is the place a non-transportable vessel. That is why when something moves in something that is moving (like a boat in a river), it uses what surrounds it as a transportable vessel rather than as a place. Place *per se* is meant to be unmoved, as would be expected, since it is a limit, and coincides with **[119]** its content (their limits, after all, coincide).[171] So from all these [arguments] it follows that place is the unmoved and primary limit of the container.

[Chapter 5]

119,3 (Ch. 4, 212a21-8)[172] The principal reason that the centre of the cosmos and the extremity of the heavens (i.e. of the motion of the whole [cosmos])[173] are thought by everyone to be in a strict sense down and up respectively is that each of them is unmoved: the one in reality, the other in appearance.[174] Also, those bodies that are light and those that possess heaviness have their natural movement in these directions: i.e. what is carried upwards is light, what is carried downwards is heavy. Now earth, water and air create the actual centre, plus <what> is beside the centre,[175] as the downward place, since each heavy body is contained by these elements: either by all of them (as with stones protruding from rivers), or by two of them (as with walkers or swimmers). But what creates the upward place is the extremity of the circular motion, and anything related to it.

119,12 (212a31-2) So the body that has a body as its external container is in place; the one that does not, is not. That is why the Earth, by remaining stationary, is in place (for externally it has another body, the limit of which contains it), whereas what moves in a circle is not in place (for nothing is external to it); in fact, not even if it were air or water, but had nothing external to it, would it be in place.

119,17 (212a31-b3)[176] How, then, do the heavens move when they are not in place? Do they both move and not move in different ways? They do not in fact move in their totality, since they do not even change their whole place (*that* movement involves wholes, like those that change position in a straight line), but they move in a circle (*that* movement involves parts, and they have a place).[177] In other words, each sphere that is carried round in a circle alters the surface of that sphere that is next to it. Thus it is also that [sphere's] place, since they are contained by one another. Similarly, while the All (i.e. the whole [cosmos]) is not itself in place, its parts are: for some of them move in a circle, while the ones that can be compacted and rarefied move up and down.

[*Eudemus' solution*]

119,26 Eudemus addresses these [problems] in Book 3 of his *Physics*, and states the following:[178] 'Are [the heavens] themselves in place, or are they not? How is each alternative [possible]? For in general terms[179] they are not in place unless there is something outside them; for that is how they would be contained. We shall investigate this. The stars, and everything inside the outermost body, are within that [body's] limit, by which it does its containing. Things that are in something in this way **[120]** are said to be in place, but what the parts are in we also speak of as "the whole". While [the heavens] would be in place in this way, they are somewhere in still another sense. That is because the whole is in the parts, and because being somewhere has several senses.'[180]

120,4 (212b3-8; 11-12) To make it clearer how we do, and do not, speak of the heavens as being in place, let us pick up the discussion from a little earlier.[181] For, as was stated, some things are in place potentially, others actually. Now the parts of a continuous body that is in place are so potentially, in that they can also be divided; for they are potentially in place because they could actually be divided, and instead of being continuous could be in contact with one another, as in a heap. But what are actually [in place] are the continuous [bodies] themselves, when they are contained by something, and also their parts, when these are divided. Again, some things are in place *per se*, like bodies that can change in respect of motion, or of increase and decrease); others incidentally, like colours and the soul. The latter, like things that exist in potentiality, will never *per se* enter a place, nor are they separable from their substrates.

120,15 (212b8-12) So how are the heavens in place? Are they so in actuality? But they are not contained by anything on the outside. But are they so potentially? But then they would never be contained. But are they so *per se*? But they cannot change in respect of motion, for being carried along is not the same as being carried round; instead, the former [process] alters the whole of a place, while the latter involves moving in the same place while remaining in the same place.[182] So [are the heavens in place] incidentally? That is Aristotle's meaning, and his commentators agree;[183] for he does speak of 'things [in place] incidentally, like the soul and the heavens'.[184]

120,21 Personally I have a problem: isn't he self-contradictory in taking 'incidentally' here as identical to 'in respect of parts', whereas earlier, in the passage in which he was demonstrating that nothing was in itself in the sense of in a place,[185] he condoned 'in respect of parts', but did not supply 'incidentally'. So perhaps here he is using 'incidentally' in a more general sense as a replacement for 'in respect of something else'. It is his frequent habit to use species instead of genera,[186] with 'in respect of parts' used in different senses here and earlier: earlier because the part is in the part, here because the whole is in the parts.

120,28 (212b12-17) How, then, are [the heavens] in place incidentally? It is because they are not continuous with everything, whereas their parts, i.e. the spheres, are in contact with one another and contain one another, and each moves with its unique movement. Because of this the parts [of the heavens] are in place. But the whole [heavens] are also in place incidentally, for as a whole they are in their parts and could not be separated from them.[187] **[121]** But not *all* their parts are in place (for all of them are not also contained), and the outer sphere is not: instead, it is in place in respect of what is on its inner side (i.e. it is in contact with the [sphere] of Saturn, and, that is to say, 'in a way' contained [by it]),[188] whereas in respect of its outer side, it entirely lacks any share in place. And its parts, which are continuous, are in the same state. For they are not in place actually or potentially (for how so, when they are inseparable from the whole [sphere]?),[189] nor indeed *per se*, but only, if at all, incidentally, and not even this unqualifiedly. That is because the outermost sphere is also not in place unqualifiedly, but [its parts] are in place incidentally in the same way that it is in place as a whole. And that [sphere] faces inwards. And the parts indeed [are in place] in this sense, since how could the All be in place in a strict and unqualified sense? For there is something outside what is in place, but nothing outside the All. For how will something be total, if indeed it is truly total, if there is also something else beyond it, which it is in? And what will that which is outside the All be? It will not be void.[190] But if it is a finite body, will that too be in place? And in what sense? For the progression [of such bodies] will be to infinity.[191] But if it is infinite [body], then the earlier arguments challenge it.[192]

121,15 (212b17-22) If the traditional belief – that everything is within the cosmos, and the cosmos is <the>[193] All[194] – is true, it would surely be irreverent to look for[195] something outside it. For it itself does not need a place, whereas all other things are in it as in a place, but not in its totality, only in its extremity relative to us, which is in contact with the body that can be affected. And this explains why the Earth is in water, water in air, air in aether, and aether in the heavens, but the heavens are not in anything else.

[*conclusion to Chapters 1-5*]

121,21 (212b22-9) We shall, then, prove what we pledged: that all the problems as to what place is are solved for anyone who works through them in the right way.[196]

So (i) place does not have to increase, since a limit *per se* does not increase at all. Instead, when the containing body gives way, either in respect of mutual replacement, or of compaction and compression, the limits also yield, so that what is increased also enters a larger place. However, if place is an extension, then place *per se* would increase and take over a larger place, and in that way a place would be in a place.

(ii) A geometrical point has no place, since a point is neither separate nor contained by anything.

(iii) Two bodies are not in the same place, since place **[122]** is not a corporeal extension, indeed not an extension at all, but a limit in which there is always an extension, yet not *per se* but along with the relevant body.

(iv) Place is somewhere, yet not in the sense of being in a place, but as the limit is in what is limited. In other words, not everything that exists is in a place, only the body that is can undergo change in respect of motion.[197]

122,5 (212b29-213a10) Also, each [body] is carried to its proper place – reasonably, since those [bodies] that are not in forced contact with one another and that contain one another are kindred (e.g. as water is to earth, air to water, and fire to air), but kindred because consecutive members of a series are very easily transformed into one another, since they have something in common. Thus by being carried to kindred bodies, they are carried to their proper places, since they are contained by kindred [bodies]. For by being in contact[198] with one another, they are also in place, and act and are acted on with one another, given that the parts of [bodies] that are conflated (i.e. unified) are not in place *per se*, nor in any way affected by one another. Also, each predictably remains in its proper place; for if it is reasonable for them to be carried to the kindred [body], it is also reasonable for them to remain with the kindred [body].[199] For they become, in effect, a part of what contains them, as air does for water, but not as a continuous part, as water does for water, but as one that is in contact *and* discrete. That is because they become one another's matter by being transformed into one another, even if not in the same way. But the latter [subject] is for later discussion;[200] right now an explicit account cannot be given, and the following suffices. If they are matter, how [are they so] for one another? Now parts that are naturally fused together and those that are in contact similarly remain within wholes.[201] But for someone who defines place as an extension, how will bodies be carried in the direction of their proper places?[202] For reasonably the limit proprietary to them belongs not to every body, but only to the kindred [body] towards which they are carried, whereas an extension is undifferentiated, and one does not pertain more than another to something that is being carried along.

122,24 (213a10-11) An account of place – that it exists, and what it is – has been given.

[Void: Chapters 6-9]
[Chapter 6]

122,25 (213a12-22) For the natural philosopher the discussion of void is next in sequence. In fact, everyone who believes that place is an extension says that void and place are identical in substrate, but distinct in definition: for the same extension,[203] when it holds a body, is also its place; but when it does not, it is a void, since by a void they mean an extension in which there is no body. So this, then, is our next discussion, the investigation to be undertaken in the same way [as for place][204]: into whether [void] exists or not, into the way in which it exists, and into what it is. But first we must establish the basis on which it is believed to exist, and not to exist,[205] and before that we must examine the standard beliefs about it.[206]

122,32 (213a22-b4) Now **[123]** those who try to show that the void does not exist do not confront the subject directly;[207] that is, they do not argue against the nature of the void that the doctrine's proponents dream up, but try and show that the void that does not exist is the one that people misguidedly believe in. That is, the conception of the void held by those who posit it as existing is that the void is an extension in which there is no perceptible body, yet, in the belief that air is not a perceptible body, they regard anything filled with air as a void. Now someone mounting the correct opposition to this doctrine does not have to prove that air is not a void, as Anaxagoras used to by twisting inflated wineskins to prove that air has strength, and with clepsydras that admit water only if the air is allowed an exit.[208] But how should the void be eliminated? By refuting its intended meaning, not what people claim. So what must be shown is that there cannot be an extension distinct and actually separate[209] from bodies – one that *either* divides the whole of body and prevents it being continuous; *or* that surrounds on the outside the whole continuous cosmos.[210] For the void can be posited in two ways: *either* as disseminated[211] in bodies, as Democritus and Leucippus claim,[212] and many others, including Epicurus later (they all make the 'interlacing' of the void the cause of bodily division, since according to them what is truly continuous is undivided); *or else* as separate (i.e. gross),[213] *per se* surrounding the cosmos, as some early thinkers were the first to believe, and later Zeno of Citium and his followers.[214] We, then, must examine what those involved with the void claim.

123,23 (213b4-12) They say, first, that there would be no change in respect of place without the void being conceded (and change in respect of place comprises motion, increase and decrease).[215] And before that they ask: how does motion *not* exist? For if what moves does not move through the void, it must move through the plenum. But if through a plenum, then a body will move through a body, i.e. a body will be in a body, as a whole in a whole, and two bodies will be in the same place. But if two, why not more? For a body is certainly not more of a plenum when it is doubled than it was earlier when it existed alone; for previously too it was equally continuous and a plenum, but still it received yet another body! So why not still another one on top of that, and again another?[216] And in this way the largest body will be in the smallest, since what is large is a plurality of small things.[217] So not **[124]** only will [the plenum] receive bodies equal to itself, but *ipso facto* ones that are unequal too – for equals are multiple and unequal when they are created by being divided in successively different ways.[218] Therefore[219] if we are to preserve movement, there must be a void.

124,3 (213b12-14) It was by yielding to the latter argument that Melissus, since he did not believe in a void, claimed that the All did not even move.[220]

124,4 (213b18-20)[221] Obviously if there is no void, *a fortiori* there is no increase,[222] since increase must come about when nutriment is assimilated everywhere by the body that is increasing, and this would not come about unless [nutriment] were totally pervasive. But a body cannot go totally through the body except by our inserting some void in bodies.[223]

124,9 (213b14-18) This, then, is what they attempt on the basis of movement, but [they argue] along other lines that when some bodies clearly coalesce (i.e. are compressed into a smaller volume) they remain attached to the same substance.[224] For example, casks obviously receive an equal amount of wine, both *per se* and as inserted into wineskins to accompany the skins, because there are some void spaces in the wine into which the latter coalesces, i.e. is compacted, when squeezed out by the force [of the additional wine].[225]

124,14 (213b21-2) They offer as further evidence the business involving the ash that receives as much water as the vessel does when empty.[226]

124,15 (213b22-7)[227] And, in general, [they argue that] bodies would not be divided or separated from one another unless the void were interstitial,[228] and so prevented them from all being unified with one another.[229] This is in fact the causation that the Pythagoreans add to the void[230] when they say that, by forming an external envelope, it is also (I know not how!) 'inhaled by the All', since the single nature of numbers is also divided – indeed there would not even be numbers unless the void separated the units.

[Chapter 7]

124,22 (213b30-1) So this is what those who posit the existence of the void say is the case, yet, as noted, those who confront this doctrine do not argue against it in relevant terms.[231] So let us state[232] whether it contains the truth. The starting-point in every inquiry of this sort is to examine what the proposed subject signifies: i.e. if it signifies a single thing, to see whether this is possible or not; but if more than one thing, to see which ones are, and which ones are not, possible; and if it is nothing that can be signified at all, to demonstrate that the expression has no subsistence.[233] So let us learn from the actual people who posit the void what they believe that this term signifies.

124,29 (213b31-214a6) Some will speak of it without qualification as 'a place in which there is no body',[234] this being equivalent to '[a place] in which there is nothing' (they believe, that is, that everything that exists is a body). If they do speak of void as 'that in which there is no body', and if, according to them, every body is tangible, and everything tangible **[125]** possesses heaviness or lightness, then, by syllogistic inference,[235] the void is discovered to be 'that in which there is nothing light or heavy'. So the consequence for those who understand it in this way is that they are saying that the point too is a void: for there is nothing in it that is heavy or light! For if they also add 'place',[236] they gain no advantage, as long as they do not determine whether they are speaking of such a place as one in which there is an extension that can receive a tangible body.

125,6 (214a6-11) This is why others will add 'extension' and say 'the void is an extension not filled with a body perceptible in respect of touch'. So *they* will certainly evade the [mathematical] point, but behold something else more problematical if this extension took on colour, sound, or some other quality apart from heaviness or lightness. Will they still speak of the extension as void, and will the void be perceptible or not? And how will it receive what is signified by 'void' (given that it is not void)? Still, it is not filled with body that is perceptible in respect of touch. But isn't the assumption impossible – that *a body* should be neither light, nor heavy, nor tangible at all? (That, after all, is what the divine [body] will be demonstrated as!)[237]

So if something like this were in a place, the place will be void, even though it holds a *visible* body!

125,16 (214a11-16) Now to avoid this calamity too others go on to explain the void as an extension in another sense: as 'that in which there is not actually a this <something>,[238] i.e. in which there is no body at all, and no corporeal nature, either tangible, or perceptible in respect of any other sense'. Yet it follows that they are also describing the void as matter, for matter is not in actuality any body. (The same people also make matter place.) Now they certainly have a better understanding of body than the others, yet even so someone could fault them in general for also making void matter; for matter is not separate from physical objects, except in definition,[239] whereas they are investigating the void as actually separate, and it slips their grasp.[240] Nevertheless, not even according to their actual doctrine can the void exist.

125,26 (214a16-21) So that now completes our examination of the argument [for void]; for when we were proving that place was not an extension existing separately from bodies, we were jointly demonstrating that the void too could not exist.[241] For if there can be no extension without a body, then neither is place an extension that can be separated from bodies, nor indeed an extension that is actually separated from bodies. For void and place, as we have said,[242] will be only conceptually distinguishable for those who posit [place] in such a way that we conceive of the place when it has already received a body **[126]** (for it is the place of some [body]), but of the void when it has not yet received a body, given that being naturally separate extensions of bodies is equally a property of [place and void] according to them. So those [earlier arguments] perhaps sufficed, yet what must also still be demonstrated here is that no such void can exist either *en masse*, or as intermingled with bodies, as more tenuous bodies are with more compact ones,[243] as people who believe that void is air demonstrate for it;[244] they do not even realise that the void is not meant to be a body, but a body's extension. After all, the reason that those who posited the void as an extension thought it just as essential as place was that there always had to be *some space* capable of receiving bodies.

126,10 (214a21-32) Both those who say that place is an extension beyond the bodies that enter it,[245] and those who [say] that the void is like this, introduce both [place and void] and establish both by launching out from the same [premise][246] (namely, change involving bodies),[247] and posit the void as the cause of change, not unqualifiedly, but as a cause in the sense of a specific space in which bodies change.[248] This would also be just the same[249] as saying that the void as place is a cause; otherwise they are saying that a body goes through a body, along with all the other claims made earlier.[250]

So since they believe that the void is above all established on the basis of change, let us first examine[251] just this claim. In general, then, it is implausible to think that the void is the cause of *all* change, and they are not saying this[252] (for alteration does not also require a place, though Melissus[253] missed this), but neither [is it the cause] of change in respect of place (for bodies can exchange positions with one another, and thus be changed in respect of place, by altering their mutual positions, without requiring a separate void extension).[254] For example, we see that water, without adding to itself an extension or place, is still not prevented from undergoing change when its parts replace one another, as in whirlpools. So why is the void required in this case to prevent body going through body? For the place that this part here occupied in the water earlier is the one that another [part] now occupies.

126,27 (214a32-b3) Now, secondly, they said: If bodies are compacted, there will be a void; for it is into this that they coalesce.[255] This too is false, for they are compacted when a tenuous body is squeezed out, i.e. in effect extruded,[256] just as water, air and earth are compressed when air, fire and water respectively are squeezed out of them. For there is always something tenuous mixed in with what is more compact, which is why fire is not subject to compression. **[127]**

127,1 (214b3-9) Third, they say: if there is increase, there is also an extension,[257] given that a body does not go through a body.[258] Now if by 'increase' they mean *every* development in physical magnitude, their argument is false and implausible. A vast number of bodies increase into a larger volume without anything even being added to them, as with water when it is transformed into air. Here the volume is increased, but a body is not added. But if it is 'increase' in the strict sense of what is derived from nutriment,[259] they are trapped by their own wings, not by others,[260] in the matter of their argument; in fact the very theory of increase totally disables itself. For to solve a standard problem by trying to introduce the void, when it obviously achieves nothing even when introduced, undermines its purpose. For while intended to preserve increase by escaping the problems applied to it, it safeguards them instead of guarding against them. For if nutriment is assimilated by the rest of the body by passing through the void, look[261] how this can be made into a problem: i.e. *either* the addition to the body occurs not everywhere but only at the void spaces, and there is no increase; *or* if everywhere, then *either* the body must be void everywhere, *or* a body must go through a body, *or*, if both [consequences] are absurd, then conversely *either* the nutriment is void, *or* there is no increase at all. Do you think that those who introduce the void have made any progress with the problem rather than flooring themselves[262] by saying that because of increase there must be a void, while failing to preserve [increase] even if we accept [the void]?

127,20 (214b9-10)[263] The argument concerning the ash[264] also awaits the same [refutation]: for *either* the water poured into the *kylix*[265] on top of the ash must be less [than the vessel's volume] (since the void spaces accompanying the ash are also less than the ones there previously),[266] *or*, if [it is a volume] equal [to that of the vessel] (this is how the problem is stated),[267] then the whole of the water necessarily goes through the whole of the ash – something they quite rightly reject. But this is not even a demonstration at all: for they do not demonstrate how a void exists, but instead confront a problem that they themselves do not even rectify.[268]

[Chapter 8]

127,28 (214b12-13) How the preceding [problems] should be solved will have been discussed even apart from the void,[269] but here we shall again investigate the void since it is believed to be the major cause of movement. And since some mingle it with bodies, while others separate it *en masse*, we shall first examine whether the latter type of void (the separate void, I mean) can be the cause of movement.

127,33 (214b13-19) To say that a void is the cause of movement is absurd **[128]** when we see that the elements' own nature is the cause of their movement. That is, each has by nature a unique motion: upward for fire, downward for earth. But how can the void be the cause of upward and downward movement? For it is undifferentiated everywhere, so that when bodies are positioned in it, they must either be carried everywhere, or remain stationary everywhere, and the void be <no> more[270] the cause of their stability than of their movement.[271]

128,6 (214b19-24) This argument also applies to those who make place a separate extension: for what they must show is how each of the natural bodies will move to its proper place, or how it will remain stationary in it. By saying that [place] is the limit of the container, we (as already stated)[272] are committed[273] to the kinship [between body and place], whereas the extension is undifferentiated everywhere. For what difference will there be between up and down? In fact, since bodies are differentiated by whether they are up or down, the limits by which they contain the ones that are kindred also differ (for we do not make limits *per se* places, like the mathematical surface, but like the one in a natural body, which is also inseparable). But those who say that [place] is the extension simply intend it to be separate from natural (i.e. uncompounded) bodies: thus the one in fire will be no different from the one in water, since it is not in any way involved with body, but exists *per se*, as also does the void, if it exists, for, as has been stated,[274] those who divert extension to place regard the void as simply place. So if this extension does not cause movement, then the void too is similarly non-causative. After all, they jointly import one another, and equally they jointly eliminate one another.

128,21 (214b24-8) Aristotle adds that saying that place is an extension entails putting its parts *per se* in place too. We anticipated in the treatment of place;[275] why, then, should we repeat ourselves?[276]

128,24 (214b28-215a1) The consequence of saying that there must be movement if there is void is quite the opposite, once it is realised that nothing can move if there is void. What follows for those who posit the void is just the same as for those who say that the Earth is at rest because of the equilibrium of its container:[277] i.e. there will be no movement one way rather than another, for [the void] *qua* void is undifferentiated.

128,28 (215a1-14) Next, since all movement is either forced or natural, then where it is forced, it must first be natural; i.e. the former is subsequent to the latter, and a deviation, since **[129]** it would not be forced (i.e. contra-natural) unless it were opposed to the natural kind.[278] So if we demonstrate that natural movement is not preserved in the void, clearly no other contra-natural ones will be preserved either.

But still, how will it be natural, when there is no differentiation in the void, which is also infinite? For insofar as they say that it is infinite, they entirely eliminate up and down, and centre and extremity, towards which the natural bodies naturally move; while insofar as it is void, they remove *every* differentiation from it. That is, the void is just as undifferentiated as *Nothing* (in fact, Democritus speaks of the void as something non-existent and a deprivation),[279] so that if natural movement needs differentiation for its container, and this does not exist in the void, then, without it,[280] there would not even be contra-natural movement, nor any other of the subsequent ones.[281] So if the void exists, movement vanishes,[282] and the claim made[283] is therefore true: that if there is going to be movement, the void must be eliminated – not introduced.

129,14 (215a14-19) So while it would suffice to demonstrate from natural movement not being preserved that contra-natural movement is not preserved, let us also discuss the latter independently.[284] Every contra-natural movement, that is, occurs *either* when the mover is present and applies force, *or* when, as in throwing and archery, the one who imparts the source is at a distance (i.e. not in contact). But neither [situation] is preserved [by the void]. In fact throwing would even lack a plausible explanation.[285] For, as it is, movement occurs when the one who throws (i.e. supplies the thrust) loses contact *either* because the air ahead of the body that is being thrown is replaced by the forward rush, and gets behind [the missile] to supply a thrust[286] until the rush is reduced or overwhelmed; *or* because the air supplies thrust, in effect thrusting itself together with[287] the missile, while the air behind flows forward in a mass along with it. For since [air] is easily moved, if it just gets started, it will advance a considerable distance by maintaining the [initially] imparted movement,[288] and by causing a movement faster than the motion of the body that was thrust forward – [a motion] by which it is naturally carried towards its proper place.[289] None of this can be described in the void,[290] in fact not even when what applies force (i.e. what in effect thrusts or attracts)[291] is present. **[130]** For how does what applies force itself move? Naturally or contra-naturally? In other words, the same absurdity will again apply.

130,2 (215a19-22) Furthermore, nobody could say why something once it moves will stop anywhere.[292] Why here rather than there? But if there, why will it not come to rest everywhere?[293] For it must either always stand still, or always be carried along. But perhaps something more powerful and stronger will in some way impede the motion?[294] But the argument will revert to the following:[295] is this more powerful thing in fact stationary, and why here rather than elsewhere? Or it moves – and how does it move, or how will it be an impediment?

130,8 (215a22-4) Also, they themselves say that nothing moves through the plenum (for *it* does not yield), but through the void, in that it lacks resistance, and is no impediment. For there is also more movement through air, since it is has greater proximity to the void, but less through water, and none through earth. But in *infinite* void one [part] is not more yielding than another, so that things will be carried identically in all directions. Also, the following must be put to Chrysippus and his followers: 'Why will the cosmos not be carried to infinity by being carried towards every part of the void identically?' In other words, 'why do they want to stabilise it *here*? Let the "cohesive *hexis*" suffice for non-dispersal, but what would make the whole cosmos plus the "cohesive *hexis*" remain *here*?'[296]

130,18 (215a24-29) Obviously there is also no movement in the void because faster and slower in movement are eliminated – a self-evident fact that does not even need arguing. After all, we see that there are two explanations for the unequal speed of bodies that are homogeneous[297] and that have the same trajectory (that is, in an upwards or downwards direction), assuming that their shapes are also identical.[298] It is *either* because what they move through differs (e.g. if one is carried through water, the other through mud; or one through water, the other through air; or one through something that remains stationary, the other through something moving in an opposing direction by which it is pushed back and so moves more slowly); *or else* it is because the actual heaviness of the objects being carried along is unequal. For if everything else is identical, the heavier one will move at a greater speed (e.g. if spheres of bronze or silver both move through air.)

130,28 (215a29-b12)[299] But set this distinction [in heaviness] aside, since it will get a special discussion.[300] What must instead be investigated is the outcome when the bodies through which the moving object is carried vary in degrees of tenuity and compactness. **[131]** For even if the same[301] weight[302] also has the same shape, it will go through what is more tenuous at a greater speed than through what is more compacted, and I am referring to an *equal* extension (water being not as easily divided as air.)[303]

For example, take a bronze sphere and let it move through air and through water along a line a stade in length. Of these [mediums] the sphere will move more quickly through the air, and do so to the extent that the air is more tenuous than the water. There will also be a proportion between the time-periods in which the sphere moves and the bodies through which it moves. That is, as water is to air, so is the time-period in which it moves through water to the time-period in which it moves through air. That is because air is more rarefied (i.e. more tenuous) than water to the same extent that the one time-period is less than the other, and the proportion is identical even if you start in reverse order : i.e. as one time-period is to another, so is one body to another.

131,11 (215b12-216a13)[304] With this established, it is clear that the void has no ratio to the plenum by which it exceeds it in terms of lesser density, for it is not even a body at all. But if the [ratio] of the body to the void is the same as the time-period in which there is movement through a body over the same spatial extension[305] to the time-period in which it moves through the void, and if the body has no ratio to the void, neither will one time-period have a ratio to the other. But there *is* a ratio between all finite time-periods. Therefore the time-period to which there is *no* ratio is not even a time-period. Something will not therefore move through the void in a time-period. Therefore it will not move at all, given that every movement is in a time-period.

To make our claim clearer still, assume that the same weight moves through very tenuous air in one hour over a distance of one stade. Now in what time-period will it move through the void over this stade? If it does so in half the time-period, then one time-period will be double the other, and the air will be twice as compacted as the void. And if in 1/3rd of an hour, 1/10th, or 1/10,000th part, the absurdity is identical. That is, you will not be able to find *any* ratio between the body and the void, but the latter will exceed *every* [quantity] identified.

131,26 (216a13-21)[306] If, for example, someone forced the weight to move through the void as well in a time-period, behold still another absurdity. For let it move through the void over the length[307] of a stade in one hour. Now it will move [over that length] through air in a longer time-period (make it two hours, for argument's sake).[308] So clearly it will move over some part [of the air] in one hour (let it move through a half-stade of air in one hour). Therefore it will move through *both* a plenum *and* a void in an equal time-period. While this is just the immediate absurdity, there will be still another more absurd one: i.e. one plenum will have the same ratio to another (the half-stade of air to the full stade) as the void stade does to the **[132]** stade full of air, if the time-period (i.e. one hour) for the movement in both the stade-length void and in the half-stade of air is also [in the same ratio] to the time-period (meaning two hours) for the movement in the air that is a stade in length.

[excursus: supplementary arguments]

132,3 These, then, are the additional problems that Aristotle raised in strictly demonstrative form for those positing something's movement through the void. But what he adds to them is not in all respects true, and would cause confusion by the decidedly unclear way it is stated.[309] So for anyone who has the leisure to spare, what follows will be culled from the commentaries.[310]

132,7 Since the consequences of positing the void are so absurd, the basic assumption from which they follow must be eliminated, and the explanation for the absurdity it entails is obvious. Since every movement occurs in a time-period, necessarily there is also a time-period for movement in the void. But there is a ratio between all time-periods, if they are of finite length. Therefore the [ratio] of the void to the plenum will also be identical to that between the time-periods in which something has moved an equal distance through the void *and* through the plenum. So the explanation along these lines of movement at unequal speeds is undermined for those who posit the void, and the [explanation] for the difference between [the mediums] through which things are carried is similarly different; for one [part] of the void cannot remain stationary, while the other moves in the opposite direction – for what movement would there be for void?

132,17 And let us see if they even preserve the [explanation] familiar to artisans. After all, when smiths, ironworkers, goldsmiths, tanners, sailors, fishermen, in a word, those whose labour involves the sea, hold up scales, or release a hook, or put down anchors, some of these objects turn out to be carried downwards faster, others more slowly. They are not stuck for an answer if you inquire as to how a ten-talent anchor and a three-talent one with identical shapes are not carried through the same sea to an equal depth in an equal time-period. Instead, they notice that the one is forced to the depths more quickly, the other more slowly; they will, after all, I think, say with a chuckle[311]: 'Ten is heavier than three!'

But the geniuses [who propose the void] will not have this explanation to offer. Instead, when spheres of lead and of pumice-stone[312] move in the void, they will say either that they move at equal speeds, or else what rationale will they be able to offer for the lead moving faster? In plena, that is, the heavier thing is necessarily carried along faster; for by its power it causes more division in the underlying substance, be that water or air, through which it is carried. But in the void, by contrast, what greater speed will lead have than the cork? For the void yields to both in the same way; rather, it does not even yield! In fact this [void] is superfluous: relative heaviness and lightness are certainly eliminated, and everything will move at an equal speed. **[133]** Therefore[313] if these consequences are impossible for their hypothesis,[314] what causes them must be rejected.

133,2 But [the proponents of the void] preserve only the difference in movement as it applies to the *shapes* of things that move. For inquire as to why a flattened piece of iron or lead lies on the surface of water, but not one that is bulbous or elongated, even if it is much smaller.[315] We can say that because flattened objects occupy a large amount of air or water on which they ride,[316] they are not carried downwards quickly, since the substrate [of air and water] is also not easily dispersed by the shape's being flattened out. But things shaped into sharp points[317] occupy a small place, which is why the air very quickly fissures whenever they are on a trajectory: for being more pliable, it is more easily dispersed. As to whether the same claims can be made by [proponents of the void], I do not know. For identical questions must also be asked about the atoms: i.e. if they are carried at unequal speeds in the void, let them tell us the cause; but if at equal speeds, let them show how they will overtake one another, or how they will be intertwined to generate something else.[318] Clearly, then, the void eliminates rather than establishes movement.

133,15 (216a26-b2)[319] And leaving movement aside to examine the void *per se*, the so-called void would clearly be devoid of meaning! For since it has been demonstrated that a body does not go through a body when one body is inserted in another body, the pre-existing one must be displaced by the incoming one, and to an amount equal to the volume of the inserted [body]. And this is immediately clear in cases such as that of a stone cube being tossed into water, for the volume of the water that will be displaced will be equal to that of the cube. But in some cases [displacement] is not entirely clear to perception, yet comes about similarly and is detected by certain devices. With clepsydras, for example, when water flows into them when they are full of air,[320] the reeds of flutes or trumpets,[321] if applied to the openings, expose the passage of air, since it makes a noise in pressing against the instruments. But if you granted [the air] no exit, and forcibly inserted a second body, it is *either* compressed into itself by being transformed into a smaller volume (i.e. made compact by being contracted) so that the vessel receives from the inserted body as large a [volume] as the compression of the original volume produced through contraction; *or* if you forced in[322] a second body, it would soon break the vessel. So it is self-evident phenomena like these, also displayed by artisans on a daily basis, to which the theory of the void cannot lay claim, or is unaware of.

133,31 Now what will they say when a body is inserted in a void?

That an equal volume of the void is displaced? **[134]** Isn't that ridiculous? That instead [the void] remains? And isn't it astonishing that an equal extension of the void passes right through the cube? For if the void does not exist at all, not even we have anything else to say; but if it has any natural reality (i.e. is extended in three directions), how will it pass through an extension like this?[323]

134,4 (216b2-12) What is equally absurd is that if the water is also not repositioned by the stone cube, then neither is the air – instead they pass right through it. For they will not be impeded from going through one another, i.e. from entering the same place, because they are hot and cold, light and heavy, plus the other affections, while the void [will] not be [impeded] because it is deprived of all such [qualities]. For it was already stated earlier[324] that being in place exists for bodies only with respect to their extensions, and these, even if they are inseparable from the other incidental properties, do still have a different being as actual extensions – i.e. the stone, when white, black, hot, cold, light or heavy, has to occupy a place of a specific volume only when that is the size of its extension. Yet because of its heaviness a body moves into this place, and when it is in a place, it is in a place in respect of its volume. In fact, it is to this [volume] alone of the properties of the cube that the magnitude of the place will also belong – not to heaviness, lightness or any other affection.

Now if you also hypothesised for the sake of argument a stone cube separated from its other incidental properties, and existing exclusively in the extension, the space it will occupy will still be equal to the one it occupied when also accompanied by the affections. Therefore in the void too it will occupy an equal [volume of] void, both when conceived of with, and without, the incidental properties. So in this case what will be the difference between the body of the cube and the equal [volume] of void and place? And if there are two such things in the same place, why not even more – even infinitely many?

134,24 Now also look in recapitulation at the necessity in the [preceding] demonstration.

(i) *If a body is in [the void], then an extension is in an extension (for the void does not withdraw, and being in place belongs to the body only in respect of its extension).*

But (ii) *if an extension is in an extension, then a body will also be in a body; for since bodies are in place only in respect of their volumes, and not in respect of any other of their properties, then if the volume can be in a volume, the other incidental properties certainly do not prevent a body*[325] *from being in a body.*

But (iii) *it is absurd for a body to be in a body.*

Therefore (iv) *a body also cannot be in a void.*

In other words, the body in a void produces a volume in a volume, but what kind of body in a body [produces] a volume in a volume? And so in this way the void stands refuted as pointless, if it is **[135]** introduced for the purpose of receiving bodies, yet cannot itself receive a body.

135,2 (216b12-16) Here's further clarification. If in each [body] the extension that is separated in definition from its affections is identical to the void, and if bodies move together with their unique extensions when they move and change position, why do bodies need other such extensions? For if, insofar as each has an extension, it needs another extension, surely the void will need another one, and the latter another,[326] and surely this will advance us to infinity?

135,7 (216b20-21)[327] From this [discussion] it is clear, then, that there is no separate void.

[Chapter 9]

135,9 (216b22-5) On the basis of rarity and compactness[328] some think it obvious that there is a void. Given that compacting involves the same body being contracted into a smaller volume, and rarefaction its being released into a much larger volume, there must, [they argue], be disseminated in bodies a void into which the ones that are compressed contract, and the ones that are rarefied expand. But eliminating the void not only eliminates rarefaction and compacting, but *eo ipso* all change too. In fact, when anything changes in respect of place, the juxtaposed bodies through which it passes contract to provide space for what moves through them – as with people walking through a crowd.[329]

135,17 (216b25-30 + 217a15-16, 18-20) But if the void does not exist within bodies, then the All necessarily swells when anything moves, as the object close [to the original mover] continually pushes its successor forward, even to the point that the heavens overflow into what is outside – exactly what happens in bathing pools, when a body is agitated[330] in them. Xouthos[331] claimed precisely this when he eschewed the mutual replacement of bodies. (And[332] the latter does not occur exclusively for [bodies] that change position in a circle.[333] It *is* more obvious in their case – they are pushed forward only to the extent that something follows from behind – but the same thing certainly happens with the other [bodies] too.[334]) It led Xouthos also to believe that if we are not going to admit the void, then bodies must be transformed in respect of an equal volume,[335] as, for example, if one *kuathos* of air came from one *kuathos* of water. Since how would [the volume] increase when it has no place and space into which it will expand, unless, he says, one conceded that the All swelled, **[136]** and in escaping the lesser [All] appropriated a more voluminous one?[336] For where will the All swell?[337] [Doing so] into bodies being impossible, it must be into a mass void that envelops it externally. And that bodies *are* transformed into a larger volume on becoming rarefied is obvious from vessels that break when the water (or fluid) is turned into air, as with jars of vaporising honey. But where will the space be for the larger volume – unless the *container* gives way by being squeezed into the void spaces?

136,7 (216b30-3) Now if they say that the void is disseminated in bodies in such a way that in something rarefied there are separate extensions that can receive bodies, we shall eliminate *this* [void] too on the same grounds used to eliminate both place and the separate void.[338] For their splitting up[339] the void does nothing to evade the problems encountered by those who speak of the mass void, since they too are producing a separate extension.

136,12 (216b33-217a6) But if it is not separate, but also mixed in, and, as one might say, fused with,[340] more rarefied bodies, their claim might be somewhat more credible, but still, not even this is true. The first consequence for them is that the void is not the cause of every change in respect of place, but only of that in an upward direction; for if the presence of more [void] makes something more rarefied and lighter, and if things that are more like this move in an upward direction to a greater extent, then the void becomes the cause only of upward movement. (They also say that fire is more rarefied than anything else because it moves upwards more than anything else.) Next, it is not as place that they speak of the void as the cause of movement; instead, the void lifts up rarefied bodies and makes them light by being carried upwards itself, just as wineskins and nets lift things into the air by being carried upwards themselves.[341] Yet how can there be motion for a void, or a place for a void? For there has to be yet another void in which the [first] void comes to exist. Furthermore, in the case of heavy objects, what cause will they offer for their being carried *downwards*? For the void in them does not also veer downwards and attract them along with it. So what cause will you speak of? If it is the nature of the bodies, then you will find the same [nature] adequate for the other kind of movement too.

136,28 (217a6-10) In addition, if bodies move faster the more rarefied (that is, 'more void') they are, clearly the void itself would be carried along fastest of all! But in its case it is impossible to identify 'fastest', since in general time does not apply to its movement; for if there is going to be a specific time-period, then the void will also have a ratio to the plenum.[342] **[137]** But if they are going to say that the void does not move *per se* (they do say that it is also mingled with body just in order to be inseparable), then let them instruct us on whether, when bodies that are compacted contract into void spaces, they squeeze the void spaces out, or leave them behind within bodies. For if they are squeezed out, then clearly they move *per se*; but if they remain, what do they contribute to the compression, if they do not leave when bodies are compressed, but are present in them unchanged? They certainly believe that the void is particularly necessary in compressions, so that bodies can be contracted (i.e.

squeezed together) into it. How, then, can an *equal* [volume] of void be mixed in *before* the compression, and again *after* the compression? And, if this is possible, how do the compressed bodies not always maintain the same volume when nothing – neither void nor body – leaves them? How can the bodies not be compressed *ad infinitum* when the void spaces existing within them are *always* of equal volume? So if the void is removed, where does it go? And how does it move? What is the natural movement of the void? They do claim that it is upwards. So how is it in time? For the speeds are incomparable.[343]

137,16 (217a10-21; om. 217a15-16 and 18-20)[344] Since they believe that there is special plausibility in introducing the void in order to preserve compacting (i.e. the transformation of bodies into unequal volumes),[345] we shall duly state how this [process] can be safeguarded without requiring the void. Now along just the same lines we could deem the void inessential when air comes from water (i.e. a larger volume from a smaller one) by reasoning that nothing prevents air being transformed into water in the same [quantity] somewhere else, and the total volume being counter-balanced. But even so we are going to demonstrate just from our own assumptions that bodies can also be compacted and rarefied without there being void.

137,23 (217a21-5)[346] We have stated, then, that there is a single matter for the opposites: for hot and cold, hard and soft, bitter and sweet, moist and dry – for every natural pairing of opposites. It cannot come into being and cease to be *per se*, since it becomes actual from what exists potentially, and while not in itself separable, its being is successively different, as it associates[347] with the substance in respect of the latter's defining principle and is twinned with it, and, while numerically one, is clothed in[348] innumerable structures as it expands and is compressed, and is shaped and altered in every kind of way.[349] For when it is compressed, it fashions[350] something compacted and heavy, but when attenuated, something rare and light. And when it changes into hot or **[138]** cold, it does not become hot and cold by an admixture of different natures (the hot and the cold are totally distinct, but the same matter, by remaining stable in respect of its own nature, comes into being in turn by transforming the opposites from potentiality to actuality); the same applies to all pairings of opposites – colour, for example, and flavour as well as the remaining qualities.

138,5 (217a25-33) And just as by being numerically one, [matter] becomes hot and cold, not two different matters, a hot and a cold one, it also becomes large and small, while being one and the same. Obviously so, since the same part of the whole of matter becomes water at one time and air at another, and applies to different volumes at different times. That is, the air and water that come to be are not different, but are both the same natural reality. And so when the same matter[351] becomes air from water, it becomes larger, yet not by acquiring anything from outside, or through a void being admixed; and when [it becomes] water from air, it duly becomes smaller, yet without anything being removed. We, in other words, differ most from those who produce coming into being through confluence and separation, in that they fabricate coming into being and ceasing to be by addition and subtraction, and do not produce the same substrate for all the elements,[352] whereas for us[353] there is a single matter that is totally altered and transformed.

138,18 (217a33-217b2 + 217b6-8) Now just as [matter] becomes hot and cold, while being one in number (not because something different is mixed with or added to it, but because it is potentially the same [qualities]), so it also becomes both small and large by being potentially both, not by acquiring or eliminating a magnitude. Now if air, while remaining air, expands into a larger volume when it is attenuated, or contracts into a smaller volume when compacted, the explanation to be given is the underlying matter, because, by being potentially small and large, it receives each of these in succession – potentially at one time, actually at another.[354] When, for example, something is transformed from being hot to being hotter, it is not so transformed by hot things that did not previously belong to the matter now becoming hot (for then the whole of such a substrate would be hot in the same way), and not by an admixture of things that are not hot. And, as things are, when [matter] develops into heat, the whole of it develops in the same way into the whole [of heat]; for if some parts of the matter remain hot when it is less hot, others when it is more hot, it is no longer **[139]** numerically identical when it comes to receive different degrees [of heat]. Instead, something else[355] hot is added to and mixed with what pre-exists. So in these cases the same holds as for something that is rarefied and compacted: the same matter is in receipt of the same [qualities], and not of something entering from without, or exiting, as for those who introduce the void. In other words, they are unaware of the nature of matter, just like those who say of red-hot iron 'Iron received the fire' rather than 'The matter (i.e. the iron) was transformed into fire'.

139,7 (217b2-8) Still another example can be used, which also accords with the fact that the parts of the circumference of the circle are also all equally convex, and that when the same circle is reduced to a smaller volume from a larger one, with the circumference converging equally from all sides, the parts then become more convex – but not by becoming convex in [parts] that in the larger circle were straight rather than convex, but by [parts] that themselves were previously convex becoming more so. For in general 'more' and 'less' are not used for something involved in an interrupted process,[356] but for[357] total intensification or remission, as with increased heat, sweetness and paleness.

139,16 (217b8-11) Similarly in the cases of smallness and largeness, matter is not extended by taking on another magnitude, or contracted by shedding one, as those who use the void as an explanation claim. Instead, the same substrate previously attached to a smaller extension is later enlarged to a greater one, and in reverse for compression and contraction, as was just shown for the circumference of the circle. So what becomes compact and rare is identical, i.e. there is a single matter for both, and rarity and compactness do not exist by the addition and deprivation of void.

139,23 (217b11-20) What is compact is heavy, what is rare is light, the latter moving upwards, the former downwards. So if void is not needed for rarity to exist, neither is it needed for upward movement; and if not for compactness, then also not for downward movement. Compactness is related to being heavy and hard in the same way as rarity is to being light and soft, but not in all cases; i.e. not everything is harder the heavier it is. There is, for example, a discrepancy in the cases of iron and lead.

139,29 (217b20-8) On the basis of these arguments there is obviously no void existing separately (whether *en masse*, or within rarefied bodies), or **[140]** indeed as intermingled and, in effect, in potentiality.[358] But someone wanting at all costs to apply 'void' to the cause of movement for bodies, and of their being in any way heavy or light, would be saying that the void was matter, for *it* can receive rarity and compactness, from which heaviness and lightness follow, along with hardness and softness: of these, the first [pair] cause motion (i.e. movement),[359] the second the presence and absence of an affection, in other words, alteration. So this is the way that things are settled regarding the void.

[Time: Chapters 10-14]

[Chapter 10]

140,8 (217b29-32) The next discussion after this is an investigation into time. But before that it is good to work through the problems concerning it – first, whether it exists; then next, what it is – even via non-specialist arguments.[360]

140,10 (217b32-218a3) That [time] either does not exist at all, or exists barely and indistinctly, might, then, be suspected from its consisting of what does not exist; for some of it has been and does not exist, while the rest is going to be and does not yet exist – and it is of these that the day,[361] month, year, as well as time that is infinite and unending, consist (for the now is not even a part of time, but a limit).[362] But something consisting of what does not exist necessarily does not exist.[363]

140,15 (218a3-8) Also, as long as everything that can be divided into parts subsists, then in all cases either some of it (i.e. of its parts) subsists, or all of them do: all do in the case of the line that is one *pêkhus*[364] long, some in the case of a war, procession or contest.[365] But if time can be divided into parts and is continuous, none of its parts exist at all, but some have existed, while others will – and I stated above[366] that the now is not a part; otherwise it would itself also be divisible into parts, would measure out[367] the whole, and would have some extension[368] (that is what the parts of things that are divisible into parts are like). But time does not consist of the nows, but of the past and the future. The now is not, therefore, a part of time.

140,23 (218a8-16) This means that we must also inquire comprehensively and separately into the way in which the now that divides the past from the future subsists – is it by persisting that the now becomes one and the same, or does one [now] succeed another,[369] so that the one ceases to be as the other comes into being? For it is totally impossible for the nows to be simultaneously multiple, nor can any other parts of time **[141]** coexist with one another as simultaneously multiple, unless we assume that one of them resembles a container, the other its content, as in our saying 'the month and the new moon are present'.[370] But this interpretation cannot apply to the now in its exact sense;[371] for it is not the now as both content and container; i.e. as what is both longer and shorter. Hence the earlier now must cease to be for the next in succession to come into being.

141,6 (218a16-18) So since everything that ceases to be does so in time, the now too will cease to be either at itself, or at another time. But if at itself, then the same thing will consequently be and not be, and come to be and cease to be, simultaneously – just as if this log here were said to cease to be 'at itself'[372] and so to be transformed into itself. But if [it ceases to be] at another time, is that also a now, or different from a now? For if [it is a now], then two nows come into being simultaneously: the now that is ceasing to be, and the one at which it is ceasing to be. But if it is different, then it will be either the past or the future. But [the now] cannot cease to be in either: not in the [past], because it is anticipating the coming into being of the now (it will therefore have ceased to be before it comes to be!); nor in the future, because then the now would never cease to be, but would always be about to[373] cease to be; i.e. always be about [to be] in the [time] in which it is ceasing to be. Now either this is the case, or time is instead the past and the future, and the now cannot cease to be in time; otherwise what belongs to [time] as partless and smallest would be coextended with what had some quantitative extension. (This sort of argument is so far rather simplistic.)

141,19 (218a18-21) But if the nows do not succeed one another, and time does not consist of nows, just as a line too does not consist of points, something still more absurd crops up. For if the now ceases to be at another [now],[374] then isn't time in all cases between itself and [the now] in which it ceases to be – as though a line is in what is between the points? And the now will exist for all this time. But the nows in the time between are infinitely many, if time, being continuous, can also be divided into infinitely many [nows]; therefore they will be multiple, indeed infinitely numerous. So if the now is one in succession to another, all these absurdities emerge.

141,26 (218a21-5) But if it always persists as the same, then, first, won't time, by being continuous and eternal, divide into *two* limits instead of one, and particularly when thought of as finite, as with the month and the year?[375] How, after all, will the now from which the month began and at which it ended be one and the same? In general, nothing divisible and finite has a <*single*> limit directed to a different part, whether it is a line, or a body's surface.[376]

141,32 (218a25-30) Secondly, if the now persists as one and the same, and if the [nows] that exist or have existed in respect of the same time **[142]** are said to exist or have existed simultaneously, then what was long ago will be simultaneous with what is most recent, the newest with the oldest – Neleus' kingdom[377] with Caesar's! For nothing will be totally past if the now persists as precisely one thing.

This does as a statement of all the arguments from which it can be established that time does not exist.

142,6 (218a30-3) What was handed down earlier regarding time is so far from in any way clarifying[378] it that any effort to instruct us on the essence of time also makes its existence equally as disputable as the arguments that we have just described.[379] For what is claimed is surely obscure and bizarre.

142,10 (218b1 + b5-9) In one case some people held that the sphere of the whole [cosmos] was time, because everything is in time *and* in the heavens.[380] First, they might be blamed for ignorance of the homonymy of the expression 'in something';[381] for being so untutored in the *Analytics* that they do not grasp the deformity of the syllogistic figure by which they hold that time and the sphere of the heavens are identical;[382] and for many other impossibilities. For example, the past and the future are parts of time, not of the sphere, and while [time] comes into being, [the sphere] exists, and while [the sphere] is a substance, [time] is not. But perhaps it is too elementary to examine this doctrine in greater detail when it is simplistic and antiquated.

142,18 (218a33 + b1-3) For these reasons others make time not the heavens *per se*, but the revolution of the whole heavens.[383]

First, according to them, a daytime will not be a time,[384] nor will a nighttime *per se*, since neither of them is the revolution of the *whole* heavens, only of the hemispheres, and so a *part* of the whole revolution. In addition, the revolution of the whole [heavens][385] is not unidirectional[386] but composed of multiple changes, and some of them contrary to one another, whereas time is one and unidirectional. Also, the movement of the whole heavens is dissimilar [to time], in fact faster and slower[387] (faster around the equator; slower around the poles), whereas time is the same for everyone.

But how is there time on the earth? Or on the sea? For they do not also complete revolutions along with the whole [heavens]. And if we supposed some manacled prisoners, such as Plato depicts,[388] living continuously from childhood in subterranean dwellings, how do they have a perception of time when they have no perception of the revolution of the whole [heavens]?

142,30 (218b3-5) Furthermore, if the universes were multiple, as Democritus says, then times too would be multiple simultaneously, and **[143]** in the same place.[389] But that is irrational, for changes can be simultaneously multiple in the same place, but that is impossible for multiple times, although the [philosopher] who assumed a 'disorderly change' even before the heavens came into being does not also combine time prior to the revolution of the whole [heavens].[390] In fact, every change is in time, whether orderly or disorderly.[391] But this [theory] does not need any extensive discussion.

143,7 (218b9-17) Instead, we must investigate in universal terms whether time can be change (i.e. a type of transformation)? Now, change for each thing is only in what changes, and wherever what changes is located (i.e. change is circumscribed by the substance of what changes);[392] time, on the other hand, is equally everywhere and with all things. Also, change is faster and slower, time is not, which is not difficult to demonstrate, since faster and slower are defined[393] in terms of time,[394] faster being what moves over a greater extension[395] in less time, slower what [moves] over a shorter one in more [time]. But time is not defined in terms of time, since there cannot be much time in a little time, or *vice versa*.

143,15 (218b17-18) But change and time differ not only in respect of quality (i.e. in respect of being fast and slow), in that this is a property of [change] rather than of [time], but in respect of quantity in just the same way. That is, the quantity of change is also defined in terms of time (i.e. the change that is in much time is long and much, the one that is in little time is little), but the quantity of time is not also defined in terms of time. For time is said to be much and little, but not by reference back to time, as in the case of change, but in another way, as we shall learn next.

143,22 (218b18-20) So from what has been said it is clear that time is not change, or any transformation.

[Chapter 11]

143,25 (218b21-7) So must [time] be investigated as not even pertaining to change at all, and so of course as not proximate to it, but instead as a stationary condition, i.e. one of rest? So how could one discover this?[396] Let us conceive, if you like,[397] of when, and in what condition, we perceive time. Perhaps that might be a means of clarification. Recall, for example, whether [**144**] you ever[398] had a deep sleep after a prolonged and continuous period of sleeplessness, such as the Poet describes with reference to Odysseus as 'sweetest sleep, like unto death'.[399] At this stage you know that, even though while asleep you have conjoined nighttime with daytime, when you wake up you still think that there has been no time in between. Legend has it that something like this happens also in Sardinia when people who are said to sleep beside the heroes wake up.[400] That is, they acquire no perception of the time they consumed while asleep, but conjoin the earlier to the later now, and in the absence of perception make it one by excluding what was in between.

144,9 (218b27-219a1) So when the now is truly one and the same, time does not exist; similarly people for whom the now is not one, though it seems to be so, believe that time does not exist. That is because in general when we are not personally transformed (i.e. changed) in any way, or are unaware of ourselves being transformed (i.e. changing), we have no perception of time.[401] This is also how night often falls unnoticed, and we continue without food whenever we take pleasure in doing something and in focusing on the task,[402] and because of the pleasure we are not made tired by change, and because we are not tired we also do not perceive change. For only those whose activity is accompanied by fatigue and pain perceive time intensely because of their perceiving change intensely. That is also why the sick regard their days as overwhelming, and in their awareness of this cry out: 'O King Zeus, how awful the nights are.'[403] From all that we have said it is clear that the perception of time subsists together with the perception of change: i.e. it is its yoke-partner, and in a precise sense they are interdependent.[404]

[*Galen on time – 1*]

144,23 So clearly there is no time without change – and without change *not* in the way Galen believes,[405] because we think of time as we undergo change (for that is what he believes Aristotle to be saying),[406] but because the conceptions of time and change are interdependent. Why did he pointlessly contest this by trying to argue against it? 'In fact', he says, 'we think of unchanging things as we undergo change, e.g. the poles of the cosmos and the centre of the earth, and still these are not accompanied by change.'[407] For he should have heard Aristotle explicitly stating: 'We perceive change and time simultaneously.'[408] For, of course, there is a vast difference between believing that time is something belonging to change because of an interdependency with the conception of change [**145**], and [believing this] because *we* think of time as we undergo change. But this fellow [Galen][409] is like this in many [instances].[410]

145,3 (219a1-10) Resuming [the commentary], we argue that: If we fail to perceive time only when there is no transformation to perceive, yet the mind appears to remain in a single and undivided state; and if we believe that time has passed when we perceive and define transformation as a given amount; then clearly there is no time without transformation (i.e. change). For even if we ourselves are at rest (i.e. not transformed in any way at all), and just think of any change, we immediately think of time too along with it. In fact, if it is dark and we are not affected through our body, yet imagine other people, or even ourselves, promenading, sailing, exercising, or making war, we also immediately think that time has passed. But not only does time follow change but change also follows time. For again when a specific time is thought to have passed, a specific change is also thought to have passed: e.g. 'Two hours have passed.' From what is this [claim] clarified? 'We have walked a certain amount'; 'We have written a certain amount.' [Time] is therefore inseparable from each [activity].

Now the first argument[411] claimed that change and time were not identical, the next one[412] that there was no time without change. So must time be posited as something incidental to change – an affection, or some property in general?[413] For these are inseparable from their substrates, yet not identical to them.

145,19 (219a10-14) So, let us resume the earlier discussion and offer this argument: Everything that changes in respect of place changes over[414] a specific extension as [it changes] from something to something; but every extension is continuous; therefore, everything that changes in respect of place changes over a continuous extension.

Now I cite this [conclusion] as the cause of change also being continuous: i.e. it is coextensive with a magnitude, and so, while not persisting, is made co-equal to the extension that also persists through this.[415] And as change is continuous because of magnitude, so is time because of change. Continuity belongs to change and time in this way primarily because of the magnitude over which change occurs, while something is spoken of as before and after *in* the magnitude. For example, this point is before, and that one after in respect of position **[146]**, as, for example, with[416] athletic judges directing the outer limit[417] of the course.[418] But since before and after are in a magnitude, before and after are necessarily in change; for change is before at the earlier point, after at the later one. But in the case of a magnitude the before and after are simultaneous (i.e. they coexist)[419] (for this point and that one are simultaneous),[420] while in that of change the before is always destroyed, i.e. it does not wait for the <after>[421] in change, unless it was before in the magnitude. Yet it certainly does not persist in the same way as [the magnitude],[422] and so continuity belongs to change because of its first belonging to the [magnitude's] extension, yet not identically, since, by contrast with change, [it belongs] to the [extension] as something jointly subsisting with it.

From what has been said it is clear, then, that the before and after are present in change, and why.

146,12 (219a14-22) But to be investigated is whether [the before and after] are identical to, or different from, change. Well, in substrate before and after both [constitute] change, but in definition differ from it; for being change and being before and after are not identical, just as before and after in terms of position are in the underlying magnitude, but being before and being a magnitude are not identical. For in explicating 'before' you also include position with respect to place, but in [just] defining 'magnitude' you will not encompass the place. So the definitions of change, and of the before and after in change, differ in the same way: for in defining change we speak of an entelechy of what changes, but a different definition must be sought for the before and after in [change]. So in substrate the father is Socrates, yet being a father and being Socrates are not identical; and in substrate five things are logs, yet their being five is also something other than the logs (five *qua* five is, after all, a number rather than logs): by the same token, before and after in change are both [identical with] change in substrate, yet their being before and after is certainly something else, and not change.

146,27 (219a22-5)[423] So we must investigate what their being is, insofar as they differ from change, and must consider how we think of the before and **[147]** after in change. So as long as we think of the whole line as one and continuous, we do not think of one part of it as before in position, the other as after, but when we use points to divide and set boundaries, we immediately make one part before in position, the other after. Similarly with change: we do not think of the before and after in it differently unless we divide and demarcate it into parts, and acquire a perception of it as belonging to one thing in succession to another. So do we use points to divide it? And how? For they do not have a position. But what position [might divisions have in the following case]?[424]

147,8 For example, I am now writing continuously, and let it be assumed that my hand changes[425] without stopping. I therefore speak of the change that has occurred as before the one that will occur. Why so? Because [the hand] has completed the one and is beginning the other. So that which can divide change has been discovered, and in its case you would identify something as you would a point in the case of a line: as the limit of one part of it, and the beginning of another. So be it. Now if anyone[426] wished to divide again the change involving the hand, will I use the same now as I also used before, or the present now, which I speak of now, though in fact it cannot be spoken of? Clearly the latter, in that it is different and is the one that is present. Of these nows which one would you describe as before and which as after? The writing is done before you can reply[427] because the one I used before to divide the change is before, the one [I used] afterwards is after. Why, then, is the one before the other? Is it because the writing is completed there on the column,[428] and now there? So if the writing is not completed, surely there would be one [now] before and another after?[429] But that is ridiculous. Yet why? So obviously the one that is before has passed, the other one is next.

147,23 (219a25-a33) So when we divide the change by using the now to segment it into a plurality, what will we do but speak of time and the parts of time? For the past and the future, of which time consists, are simply the before and after in change, and these are always demarcated by the nows and counted as successively different. Thus time, as well as the before and after, are, when demarcated and counted, identical in change, and they are demarcated only when the nows confront the mind as two: the one as before and the terminus of the earlier change, **[148]** the other as after and [the terminus] of the later [change], i.e. just as when we conceive of them as extremes and distinct from the middle. For when the mind, by recalling the now it spoke of yesterday, speaks in turn of another as today's, it also then immediately thinks of time as defined by the two nows, as if by two limits, and can thus state[430] that a quantity is fifteen or sixteen hours, like using two points to segment a *pēkhus* in length from a line of infinite length. Yet to recognise time it is also often enough for the now to confront the mind as just a single thing, but not without qualification, but only when it has the status of terminus and beginning; for that is also when before and after are perceived together: i.e. the limit belongs to what is before, the beginning to what is after.

148,10 (219a33-b3) Thus from all quarters the nature of time clearly emerges as having its being in the distinction between the before and after in change. Time is therefore the before and after in change, when it is distinguished as belonging to one thing in succession to another, i.e. when it is counted. For number does not belong completely to one thing or even to the same thing, but to what has one thing in succession to another. Time is not therefore change, but is the [aspect] of change that is counted with respect to the before and after. To sum up what has been said: time must be separately defined as the number of change in respect of the before and after.

148,17 (219b3-9) Now a sign that [time is] a number is that everything that is more or less is assessed by number [and more and less change by reference to time];[431] time is therefore a specific number. That we have necessarily added[432] 'in respect of the before and after' is clarified by the fact that changes are counted and are simultaneously multiple, as, in your case, for example, from the fact you are growing, getting hot, and advancing from place to place,[433] i.e. that you are simultaneously undergoing three changes, but it is not this kind of number of change that is time, but the [number] in respect of the before and after. There should be no surprise if in speaking of time as 'the [aspect] of change that is counted' we substituted 'number' in the definition. For what is counted is also spoken of as a number, just as what is measured too [is spoken of] as a measure. A measure, that is, also has two senses: that by which we measure, e.g. the wooden *khoinix* and the clay *khoeus*,[434] and what is measured on the basis of them, e.g. the corn in the *khoinix*, and the wine in the *khoeus*. After all, the *khoinix* we eat is not the wooden one, and the *khoeus* we drink is not the bronze one. By the same token, a number is not only what consists of the units, but also what is counted: **[149]** because the former is divided and not continuous, it would also not signify to us the essence of time, but instead just produce some end-result,[435] whereas nothing prevents what is counted from also being continuous, as with the 'spear-shaft of eleven *pēkheis*'.[436]

[*Galen on time – 2*]

149,4 We must not align ourselves with Galen in his belief that time is separately defined through itself.[437] For after fully listing numerous significations of 'before' and 'after', he says that none coincide with the definition [of time] except the one in respect of time, so that time is [defined as] 'the number of change in respect of time'. But it must be realised that the before and after in change is not the before and after because of time; instead, the before and after in time produces it, and it comes into being from what [exists] in respect of magnitude and position – the source of its also having continuity. Aristotle states this explicitly: 'The before and after are prior in place and thereby by position; but since in magnitude, necessarily in change too.'[438] But let it be granted that 'before' and 'after' signify nothing else in change than 'in respect of time', as [Galen] believes – why is absurdity the consequence of this? It is because we do not even say that time is something different from the before and after in change. But it is absolutely necessary that the definitions signify the same things as the terms [to be defined], so that he misses this in faulting the argument.[439] Hence we should do better to accept [Aristotle].

149,20 (219b9-12) And time's always coming to be as one [time] in succession to another is because of change; for change too is always one [change] in succession to another. But when multiple changes come about simultaneously (e.g. at Athens, Megara and Corinth), how do multiple times also not come about, since the now is identical everywhere?[440] It is with reference to this that difference and sameness are conceived for the now, and there is one time when the now is one, multiple times when the nows are multiple and different, i.e. the nows become multiple and different when one is identified as before, another as after.

149,26 (219b12-15 + 19-21) The latter [conclusion][441] could also be made into a problem:[442] *how at any time do such nows also become different*? For a single [**150**] nature is fundamentally manifest in both, even if one is thought of as before, the other as after – unless Socrates is different when he is in the Lyceum[443] and when he is in the Agora, as the Sophists claim.[444] Well,[445] just as Socrates is identical in substrate, whether he transfers himself here or there, but distinct in definition (for being in the Lyceum and being in the Agora are not identical), so the now is also one in nature and essence in respect of its substrate, but different in definition. For there are different definitions for the now of which Hector speaks to his pair of horses ('*now* pay me off for your fodder'),[446] and for what is *now* being written. To learn better what I am saying, we must recapitulate something from a little earlier.

150,12 (219b15-18 + 219b21-220a4) It was stated that change follows magnitude, and time change.[447] Now the first principles of these things (i.e. their capacities for producing and generating) also reasonably entail one another: the point can produce magnitude; what is moving[448] [can produce] movement; the now [can produce] time. So the now is one in the same way that a point and the stone are both one, whether they are transported here or somewhere else, yet differ in definition: i.e. the now follows what is moving, as time follows the motion, and the now is knowable only through what is moving. For the before and after in change are counted in accordance with the changing position of what is moving; for they would not be separated if what is moving remained stationary, but since what is moving reaches earlier and later places at different times, the before and after are consequently included in change too, and the now is what is counted as before and after, so that in these cases too the now that is before and the one that is after are identical in substrate, but different in definition. For the[449] existence of something itself countable and its being distinguished as one thing succeeding another supplies [the now] with its differentiation in respect of its definition. Just as what moves is more knowable than motion (for it is a this something, i.e. a substance), the now too is [more knowable] than time (for it alone fully exists,[450] i.e. is like a this <something>).[451] That the nows and time each follow one another (i.e. that the one does not exist unless the other does) is obvious. For time will exist [**151**] by the number of the nows, since the now is in effect the unit of time.

151,1 (220a4-11) And the now produces time just as the unit does number – by being identified repeatedly; and the now divides time just as the unit, though itself undivided, renders the number divided. But the difference is that the one [sc. the unit] makes for continuity, the other [sc. the now] for division. And the now holds time together and causes division as what moves does for motion. In fact motion is one and continuous because what moves is one, yet not unqualifiedly one, since, if it ceases to move, it can be like a this-something, i.e. a single body (e.g. a stone or a log), but not like something moving. For then the movement is no longer one and continuous, though what is moving is one, except when it exists as a single thing that is moving, i.e. when not at rest *en route*, for then it also divides the earlier from the later movement, and in turn holds them together, just as the point does with the line.

151,12 (220a11-18) But the difference is that you can identify the same point twice, once as a terminus,[452] once as a beginning (for its being persists even in its [fixed] position), whereas you cannot identify twice what is moving insofar as it is moving, since it is successively one thing after another. And it is another thing not because it is different from itself, but because it is in successively different things. If it is in the same thing twice, it must be stationary, but it is assumed to move[453] continuously. In just the same way the now too cannot be identified twice. So if it is impossible to identify the now twice, then in the process of being counted isn't the now like the same point <being identified>[454] twice – as a beginning and as a terminus – but in that case like the two limits (this one and that one) of the same line? But *these* [limits] bound the intervening extension of the line by being successively distinct. So the way that the nows too are counted is as successively different, one before, one after, but identifying the same [now] twice is as impossible as making time stand still.

151,23 (220a18-21) And further it is also obvious that the now is not a part of time;[455] what is moving, for example, is not also [a part] of motion, nor the point [a part] of the line. For a part of a line is a line, of a change a change, and of time a time. Of what, then, is the now a limit? Of the change or of the time (given that it was stated earlier that the now divides the change [**152**])?[456]

152,1 (220a21-4)[457] Well,[458] [the now] *per se* is the limit of time, and since time is simply what is counted in change with respect to the before and after, only when the now is identified as counted does it also become the limit of change. This, then, is why being a limit [of change] and existing everywhere simultaneously are incidental to it, and the reason is that [the now] *per se* is not the limit of *change* (for it is not only in the change of which it is a limit), but of *time*, which is the number of change. The same number can be of more than one thing at the same time, e.g. ten horses and logs. Yet time is a number not as what does the counting, but as what is counted, i.e. as time in relation to a unitary number. But still it alone assumes the uniqueness of what does the counting; for what is counted is at the same time one and the same everywhere even for multiple changes – for increase, motion, and alteration.[459]

152,12 (220a24-6) So in this way it has been adequately demonstrated that time is a number, i.e. the number of change with respect to the before and after, and clearly it is also continuous, since change is in all cases continuous.

[Chapter 12]

152,15 (220a27-32) In speaking of [time] as a number, will we not necessarily posit it as producing some smallest time too? For the smallest in number is two,[460] since there is no reduction of number to infinity, given that it is determinate.[461] Well, number is twofold: one type consists of units (including the smallest [number]), another of units not without qualification but of ones that are specific (e.g. horses or lines). For example, five lines as well as five triangles are said to be a number, and in terms of numerical quantity such things will themselves also have what is smallest (for the smallest is two lines, and by reference to it things of this type can also be counted),[462] but in terms of magnitude they will not have it (for each of them can be divided for ever). Now inasmuch as time is also something continuous,[463] it will not have what is smallest (i.e. it is not countable in this respect), but inasmuch as it is defined and, in effect, segmented by the nows that are before and after, it is countable, and does have what is smallest; two hours, for example, will have what is smallest in number, but not in magnitude.

152,28 (220a32-b5) Now time is reasonably said to be much and little, i.e. long and short;[464] for it admits both as the [properties] of continuous as well as divided things. Insofar as it is continuous, it is also long and short; insofar as it is countable, it is also much and little. Yet it is not fast and slow; **[153]** for no number is fast and slow.

153,1 (220b5-14) It is reasonable that while the same [time] is also simultaneous everywhere, times that are before and after are not identical relative to one another. In fact, the present [change] is one, while those that have been and will be are distinct. Next, time is a number – not as that by which we count, but as what is counted. <So what we count by>[465] is identical in all cases (e.g. ten, whether it be identified for horses or people), but what are counted (the ten people and ten horses) are not identical. Time too is counted with respect to the before and after, so that quite reasonably the latter are distinct.[466] Also, just as change can be one and the same again <and again>[467], so can time (e.g. spring or autumn).[468] Such is the essence of time.[469]

153,10 (220b14-24) In other words, the before and after in change is, when counted, time, but is incidental to [time], which has its being in measuring change directly, while also being incidentally counter-measured by it.[470] That is, we speak not only of much change in much time, but also of much time in which there is much change. So it is just as with all other measures too: measures are counter-measured by what is being measured. For example, it is by a *medimnos* of corn that we judge whether a *medimnos* of wheat is larger or smaller, and by a *kotulê* of wine that we counter-measure a *kotulê* of metal.[471] And what is remarkable about this happening with continuous measures, when the unitary number itself is in a certain sense counter-enumerated by what is being counted? In other words, just as we say that ten things are ten horses, so we also say that the units are ten because ten is made equal to the horses. For we might never arrive at a conception of the ten [units] if we did not identify one horse after another[472] that many times.

153,23 (220b24-32) And it is a reasonable consequence that *time* measures, and is counter-measured; for, as has been said,[473] change follows magnitude, and time change. So the status of change in relation to magnitude is the same as that of time in relation to change. **[154]** And just as change measures magnitude, it is also counter-measured by it. For example, we say that the road is long if the journey is, and that the journey is if the road is. Equally we say that the change too is much if the time is, and that the time is if the change is.

154,4 (220b32-221a9) How, then, does time measure the change? It is by marking off from it a part that will measure out the whole [change], just as the *pêkhus*[474] does to the length by identifying a specific length that will measure out the whole. What, after all, is the hour but so much change of the Sun?[475] (The month and the year too.)[476] But since change is in time by having its being measured out by time (for the being of change is nothing but its coming into being; that is what time is coextensive with, and measures out),[477] clearly for other things too being in time is having their being measured out by time.

154,12 (221a9-13) For 'in time' has a twofold sense: (i) existing just when time exists; and (ii) as we say of some things that they are 'in number'. But sense (ii) is also twofold: i.e. either being in number (ii-a) because a part or affection of number (like even, two, and odd); or (ii-b) because of being a number of <the same> things,[478] in the way that ten horses and ten cattle are 'in number'. So if time is a number, clearly what is in time is also in number. Hence the now and the before are in time and number in just the same way as the unit and the odd are in the unitary [number];[479] that is because some things are incidental [properties] of number, others of time. But physical objects are in time in just the same way as ten horses are in number.[480]

154,21 (221a26-30 + a13-19) Now there is always a number larger than the things that are said to be in number in the latter way (for their limit is fixed); similarly there is always a time of greater duration than the things that are in time in the aforementioned sense (for a time is a specific number).[481] If so, things that are in time are contained by time, just as **[155]** things that are in number are contained by number, and that are in place are contained by place. And this is what is signified in a strict sense by 'in time'.

155,3 (221a19-26) But to speak of something being 'in time', because it exists when time exists,[482] and of something being 'in change' because it exists when change exists, is no different from also speaking of the heavens being 'in a millet-seed' because they exist when the millet-seed also exists. In fact, being simultaneous is incidental to many things, but one of a pair being in the other means the one being contained by the other. However, what follows for them is that they are also simultaneous, yet it is not because they are simultaneous that they are in one another: instead, it is because they are in one another that they are simultaneous.

155,10 (221a30-b2) This [simultaneity] is incidental to the things that are in time, as is the saying that something is affected by time. For we standardly say that time wears things out, and that everything ages by the agency of time, and forgetfulness is caused by time. That is, we do not similarly say that the house comes into being and ceases to be by the agency of time, but instead that it comes into being in time by the agency of the house-builder, whereas it ceases to be both in time and by the agency of time. And [we say] that we learn in time by the agency of the teacher, but that we forget only by the agency of time; for in this second case no other cause is observed.

155,17 (221b2-7) Time *per se* is reasonably said to be the cause of ceasing to be, since it is the number of change, and change removes what pre-exists. If this claim is correct, then obviously what is always the case, insofar as it is always so, is not in time; for it is not contained by time, nor in any way affected by time (i.e. it is not altered along with it, nor does it accompany it). Hence the being of such things is also not measured by time (i.e. is not defined by it), nor does any time extend beyond their being, but, quite the opposite, their being extends beyond any identified time. How, then, is change in time? (It has been, and will be, proven eternal.)[483] Well, it is because one change always succeeds another, and is never numerically identical, but accompanies, and is altered along with, time – but in fact time [is altered along] with [change], and they need one another to exist, so that even if time does not also extend beyond change, nor contain it in that way, at least change is in time because they are harmonised with one another.

155,30 (221b7-23) Since time is the measure of change, it will also be the measure of rest (rest being the deprivation of change). We discern positive states[484] and deprivations by the same means (e.g. light and dark by the eyes, sound and silence by the organ of hearing); **[156]** thus [we discern] rest too in time, yet not in all cases, but since[485] it is in time, it is also in change. Time, that is, is not change, but is the number of change and something incidental. But if something is in something incidentally, that does not mean that it is also in the thing to which it is incidental (e.g. if something is in a day, that does not mean that it is also in the motion of the Sun). Hence what is at rest cannot be [included] in change, but can be [included] in the number of change: for example, we say that 'a tired person rested in a specified amount of time.'

But the way that time measures change *per se* is not how it also measures rest; instead it does so incidentally, i.e. by measuring *another* change. For example, night *per se* is a measure of the Sun's motion below the Earth, but incidentally of the rest taken by living things, and what is said to be at rest are not totally unchanging things, only those naturally disposed to, but currently deprived of, change. Time, in other words, is of greater duration than things that are unchanging in the sense of being at rest, and things at rest are reasonably said to be at rest in time because time belongs to their state of rest. But time will measure what is changing *and* what is at rest by the way that the one changes and the other is at rest: i.e. [it will measure][486] just how much their change and rest is. So anything that changes is not measurable by time without qualification. Even the Sun, for example, is not measurable by time without qualification, since its being is not measurable by time, only its change, insofar as that is some quantity.

Not to be overlooked is that while time could be said to be the measure of both change and rest, its number will be only of change, given that the before and after, of which time is the number, are in change, not in rest (for being for [rest] does not consist in a repositioning from something before to something after),[487] so that from what has been said it is obvious that things that neither change nor rest are not measured by time; for time is the measure only of change and rest.

156,25 (221b23-222a9) The preceding [discussion] entails that things that are always not the case are also not in time (e.g. things that cannot in any way be the case, such as the diagonal's being commensurable with the side), for these neither change nor are at rest, whereas time is a measure only of change and rest: of the former *per se*, of the latter incidentally. So everything that can cease to be and come to be, and, in a word, everything that is the case at one time but not at another, must be in time: for time, which surpasses their being (i.e. [surpasses] the [time] that measures their being) is something of greater duration. **[157]** Of things that are not the case, those that time contains are in time, the container being the past or the future, or even both: the past of what used to be (e.g. Homer), the future of what will be (e.g. an eclipse), and both of what used to be and will be (e.g. war involving Persia).[488] What [time] does not contain are things that neither used to be, nor are, nor will be [the case], and also among things that are not the case the kind whose contraries are always the case, as, for example, the non-commensurability of the diameter is always the case, i.e. this will not be in time. Therefore neither[489] will its commensurability, since it is contrary to what is always the case.

Things whose contraries are not always the case can both be and not be, and they possess coming to be and ceasing to be.

[Chapter 13]

157,10 (222a10-20) The now is spoken of in both a strict and in an extended sense.[490] First, the one spoken of in a strict sense must be discussed, and this, in line with what was also said earlier,[491] will be the one by which time is both held together[492] and divided. In fact, it holds both past and future time together (for both coincide with this common boundary), and it also divides them, just as the point does the line, with the difference that the point persists, the now does not, but is always successively different; and the [point] divides the line actually, while the [now] divides time potentially. For in general [the now] cannot even be identified in actuality (for it always anticipates anyone who wants to identify it),[493] but only in conception. So insofar as it is identified as causing division, and as one thing's terminus and another's beginning, it becomes two in definition (for being a terminus and being a beginning are not identical, even if both apply to one substrate); but just to the extent that it [is identified] as holding together, it is one in an exact sense – in respect both of substrate and of definition. It is just like the point in the case of mathematical lines: if we identify it as dividing, then in definition and conception there are successively different points; but if as holding together, then it is one and the same. Similarly the now too is in all cases one in substrate, but in definition both one and not one: one when identified as holding together, not one when [identified] as dividing, yet they are potentially identical. That is the difference from the point:[494] for both dividing and holding together apply to it, but the being for what divides and what holds together is not identical.[495] So that is the now in its strict sense.

157,30 (222a20-4) The now is also spoken of in another sense. To explain. Time, both past and future, is positioned on each side of this type of [now] and, through being near it, is itself also spoken of as 'now' by being in the environment of the now in its strict sense.[496] For example, 'So-and-so will be coming *now*' **[158]**, meaning today, and 'So-and-so came *now*', meaning today.[497] But the [events] at Troy did not occur '*now*', nor [will] the Flood [occur] '*now*',[498] yet the time up until these is continuous, except that they are distant from the present now.

158,4 (222a24-8) 'Somewhen' (*pote*)[499] is a time bounded by the present now as well as by the [now] that is before and the one that is after,[500] as, for example, 'Troy was captured somewhen', and 'There will be a flood somewhen.' For it must be delimited (i.e. bounded) in relation to the now that is before and the one that is after; for it is to these that 'somewhen' applies. Evidence of this is that we do not say in the case of [events] on which we cannot place boundaries, that we are conscious of them as [happening] 'somewhen'.

158,9 (222a28-33) But if *every* identified[501] time has 'somewhen' predicated of it, then every identified time is bounded. Will [time], then, give out somewhen? Not so, since that also means no change.[502] So is [time] different, or identical repeatedly? Well,[503] clearly change and time are both in the same state: i.e. if [change] occurs as one and the same somewhen, then time too will be one and the same; [504] if not, it will not.

158,12 (222a33-b7)[505] The now makes it clear that [time] is one thing in succession to another: for since the now is simultaneously a beginning and an end, it clearly belongs to successively different [times], since it certainly does not belong to the same [time]; for then opposites[506] would[507] simultaneously apply to the same thing. And time will *not* give out, for it is always at a beginning; and it is always at a beginning because it is also always at the now.

158,16 (222b7-12) *êdê* ('all ready'/'already')[508] is near the present undivided now, as a part of future time. 'When is he going for a walk?' '*All ready*', because the time in which it is going to happen[509] is near. And for past time there is also [*êdê* as] 'not far from the now': 'When did you complete the walk?'[510] '*Already*', meaning today or yesterday. But uttering 'Troy is *already* captured' is not our usage, because that is too remote from the now.

158,20 (222b12-16) And 'recently' (*arti*) is what is near the present now, as a part of the past. 'When did you come?' 'Recently', if, that is, the time is near the present now. 'Long ago' (*palai*) is what is remote. 'Suddenly' (*exaiphnês*) is used for what happens unexpectedly, but mostly for what crops up[511] almost imperceptibly in a little time, i.e. because [the time is] minuscule.

158,25 (222b16-27) Since everything that comes to be and ceases to be is in time, some reasonably say that time is very wise, others that it is very stupid, Simonides using 'very wise',[512] because people become knowledgeable **[159]** by the agency of time,[513] Paron 'very stupid',[514] because people also forget by the agency of time, the latter's statement being far more correct, in that all transformation naturally causes displacement. So clearly time *per se* is more a cause of ceasing to be than of coming to be, as was also claimed earlier.[515] In fact, change (i.e. transformation) *per se* causes a displacement from what pre-exists. And since some things happen on occasion to be displaced from their former essence and transformed into a different one, change and its number[516] will also become incidentally the cause of their coming into being. I mean that for things that come to be not only does time <not>[517] suffice for their coming to be, but some activity is needed through which they come to be, such as skill, natural capacity, instruction, or action. By contrast, in the case of things that cease to be, [time] alone suffices for ceasing to be, even if, as with things that putrefy in time,[518] no external cause is present. And this kind of cessation in particular we standardly say occurs 'by the agency of time'. Yet time does not even cause this, but in fact transformation of this kind[519] just happens[520] to occur in time. And the nature of each thing (i.e. its underlying matter) also causes it to cease to be, since the parameters[521] of each thing involved in[522] coming into being through the agency of Nature (i.e. in increase, culmination and decline) are determinate.[523]

159,17 (222b27-9) That completes our account of time: that it exists, what it is, and the number of senses in which 'now', 'at a given time', 'recently', 'already', 'long ago' and 'suddenly' are used.

[Chapter 14]

159,20 (222b30-223a4) With matters determined in this way, it is obvious that every transformation (i.e. everything that changes) must be a process of change[524] in time. Faster and slower, that is, are aligned with every transformation, faster being simply what is transformed into the substrate earlier. And by a substrate I mean that into which the change comes about, e.g. a place, shape or affection.[525] To clarify further: in the case of motion in respect of place,[526] when two things move with a smooth movement over an equal extension, the one that will completely cover the extension earlier is said to move faster. An identical extension (e.g. a straight or a circular line), that is, must underlie what moves **[160]**, and both [the faster and slower thing] must move with a smooth movement. For that is how the faster one will be recognised by its reaching the terminus earlier, since, apart from the difference [in speed] between each of them, sometimes even the slower one may turn out to be transformed earlier, the faster one later.[527]

160,5 (223a4-15) But whereas these [claims] are non-specialist, it can be precisely stated[528] that every change possesses faster and slower, which are the before and the after in time.[529] 'Before' and 'after', that is, are used with respect to the distance to the now (the now being the boundary between the past and the future),[530] so that since the now is in time, <the before and after> will <also be in time>;[531] for what the now is in is also the distance away from the now.[532] Clearly what we say is true: for if every change has the before and the after, and these are in time, then every change is also in time.

160,13 (223a16-b1) But why is time believed to be in everything – the earth, the sea, and the heavens? It is because it is an affection (or *hexis*) of change; i.e. time is what is counted in change, and all the [locations] just mentioned can change, since they are all in place, either totally or partially, and change and time are yoked to one another, so that whatever change is in, so too is time. But what I am speaking of is potential and actual: i.e. everything potentially changeable is also potentially in time (as with what is at rest), and everything actually [changeable] is actually in time. But if time is what is counted in change, then if there is nothing that counts, can there be something that is counted? But if nothing except a soul (and within a soul the intellect) is naturally fitted to count, can there be time if there is no soul? Now, if number is spoken of in two ways – as what is countable and as what is counted – the one (what is countable) is clearly potential, the other actual, and if these could not subsist unless there were something to do the counting, either actually or potentially, obviously there would be no time if there were no soul.

[*Boethus on time*]

160,26 Yet, says Boethus,[533] 'at least nothing prevents number existing even apart from what does the counting, just as I think there is an object of perception even apart from what perceives'.[534] But he's in error. For relatives go together, including ones that are in potentiality relative to one another, so that if there is not also something that can count,[535] neither can there be something countable. But there can be something incidentally countable even without something doing the counting **[161]**, yet not as countable in such a way that there can be change even without it being counted.[536] And the before and after in [change] are distinguished (i.e. separated) as things whose distinction, separation and number produce time. How is the preceding possible unless there is a soul? In fact, there cannot even be change without a soul; for the beginning of every change is the circular motion [of the heavens], for by its agency the affections of bodies are altered, and increase and decrease, but this is through the agency of intellect and in accord with desire.[537] And for animals change is the product[538] of the soul, and time is the number of every change; for every change has the before and after, given that not only is motion in respect of place continuous and one, but also increase, decrease, coming into being and alteration.

161,11 (223b1-10) So surely times are simultaneously multiple when changes are simultaneously multiple? Certainly, if time is the number of every change, or if the now is numerically one and the same in every [change], even if one [change] is motion, the other alteration. But the now is what is counted in every [change], and what is identified as the before and after, so that if it is identical in every [change], then it is reasonable for time to be identical in every change. For if time were incidental to changes in some other way, then it would be reasonable for it to be divided along with them (just like pale and dark are with bodies).[539] But since it is as what is counted (i.e. as their quantity) that it belongs to them, nothing prevents it being one and the same in a multiplicity. In the case of the unitary number what is counted is, I think, identical, even if the things that are counted are multiple and distinct from one another (some being dogs, others horses, but each [set]⁵⁴⁰ the identical thing that is counted). This is just how it is with changes⁵⁴¹ that are simultaneous: the time is identical, but not the rapidity or the place. For it is not by being counted⁵⁴² in these⁵⁴³ respects (meaning rapidity, slowness, or any other differentia – e.g. if the different [changes] were alteration, motion or increase) that [changes] produce time, but only in respect of the before and after, which is identical in *all* simultaneous changes.

This is what Aristotle says. As for problems that might be raised for his statements, let the plan be to inquire together.⁵⁴⁴

[*three problems on time*]

161,29 (ad 223b20-1)⁵⁴⁵ If the before <and after> are first of all in the magnitude (i.e. in the extension) to which change applies, and thus secondarily in time, how does the before and after apply to increase and alteration? These [changes] are not **[162]** in an extension (i.e. not from somewhere to somewhere), but the body is heated gradually as a whole and increases as a whole. So in these cases where will time have the before and after from, as in the case of motion from extension and position? But if it does not have the before and after from these, how will it possess their number? But in the case of increase someone might perhaps claim that there is 'from somewhere to somewhere', in that the transformation is from the smaller to the larger place. But for the cases of alteration, and coming into being and ceasing to be, what shall we say? For in their cases time will simply import before and after from itself.⁵⁴⁶ Now time either directly measures only change in respect of place, or else, if it does measure the other changes directly too, then it does not need the before and after in terms of position.

162,11 (ad 223b13-14)⁵⁴⁷ Next, if time is the number common to every change, it clearly does not have its being from change; for if it is not from this or that [change], then it is not even from change at all. For claiming that the before and after is identical in *all* changes is absurd. For what must be specifically investigated is how the *now* is the same, in genus, kind, or number. If in genus or in kind, then it is, of course, numerically multiple. But if it is numerically one, how do things that are differentiated numerically have a property⁵⁴⁸ that is numerically one? But this is how [Aristotle] himself labels time as belonging to change. For something numerically one and the same cannot belong to a multiplicity; for in this way it will become simultaneously one and not one, so that⁵⁴⁹ if the nows are multiple in the case of multiple changes, then they are simultaneous in the case of simultaneous ones – a situation that cannot even come to be conceived.⁵⁵⁰ But otherwise the time of the change is not in any way identical,⁵⁵¹ since it would make no sense for what is counted in changes to be identical in the way that it is in ten horses and ten cattle: for what is counted there is not even identical, save verbally, in that both [sets] are ten, which is our conception and observation of what is **[163]** similar in different things; but there is no substantiality *per se*.

163,1 (ad 223b13-15)[552] As to this [conclusion],[553] how in the case of time is it reasonable for it to be a conception of our soul and *not* have its own nature, as Aristotle seems to grant when he allows[554] that no time exists if there is no soul, since to speak of time as both a measure and a number is to grant a conception of this kind?[555] 'For', as Boethus says, 'no measure comes about naturally, but both measuring and counting are in fact our activity.'[556]

163,7 The preceding [claims] should be examined repeatedly and not accepted uncritically, but since[557] each measurable thing is measured by a single thing akin to it (units by a unit, lines by a line), and since time too is measurable, clearly time too would also be measured by some determinate time. What this is must be investigated.

163,11 (223b16-23) Obviously, then, this [time] will also be the measure of a specific change, since all time is the measure of change. If we identified a specific change that is going to be the measure of the other changes, clearly its time will also emerge as the measure of time. So what is *this* change? Now alteration, increase, and coming into being are not uniform throughout,[558] but motion is, and within motion the circular motion of the whole [heavens], and its number is also very easily recognised along with it.[559] This [circular motion] is, therefore, the measure of changes, and becomes a measure by being counted by a specific time, and by becoming such and such a quantity. The year, the month and the day, that is, are names of a time, and boundaries of so much change, as, for example, the hour is a time,[560] yet it defines so much change in the revolution of the whole [heavens], specifically of the Sun,[561] and by defining it makes it the measure of all changes, and not just of the changes, but also of the time that is of greater duration.[562] For [time] itself measures time: the hour is the measure of the day, the day of the month, and the month of the year, and in turn the quantity of the revolution measures the year. This is clear from direct observation too: i.e. years, months, days and hours are measures of time, and these are quantified changes in the revolution of the Sun. So it was not illogical for some people[563] to believe that time was the change in the revolution of the heavens, because the other changes are measured by this (that is, by its time). 'How much time did it take to capture Troy?' 'Ten years', that being how many revolutions there were **[164]** of the Sun.

164,1 (223b23-224a2)[564] Hence the following *cliché* too will have been well turned: *that human activities are cyclical*, that is, that they are all discriminated by time, and acquire a beginning and an end as if in accord with a specific circuit of time. Time is thought to be a cycle because it is the measure of such motion, and is counter-measured by it, and what is measured is thought to be simply multiple measures, as the stade is multiple *pêkheis*.[565]

Notes

1. Themistius omits this reference to the subject-matter of the preceding book of the *Physics*, although it is explicitly mentioned in the Aristotelian text.

2. cf. 122,24-5 (*ad* 213a10-11) for this introduction recapitulated; for similar introductions see ch. 6, 122,29-31 (*ad* 213a12-14) and ch. 10, 140,8-10 (*ad* 217b29-32).

3. See 103,30-104,1 below on Hesiod.

4. In the hypothetical 'If (p) all non-existing things are nowhere, then (q) all existing things are somewhere', (q) is the erroneous 'universal posit' (*katholou thesis*) to which Themistius is referring. The correct consequent is 'Some existing things are somewhere', since other existing things may be nowhere, or, at least, may be thought to be nowhere. Hussey *ad loc.* refers to *Phys.* 209b33-5 where Plato is cited as not believing that forms and numbers occupied place. For an elaborate analysis of the erroneous conversion in the traditionalists' conception of place see Simpl. *in Phys.* 521,15-24. For the notions of being somewhere and being in a place as trivially implying one another without ontological commitment see *Phys.* 206a2-3.

5. This can be converted into 'All bodies capable of change through locomotion exist in place'; see below 122,4-5 (*ad* 212b28-9).

6. I read *kath' hauta* (102,15) with MS Vat. gr. 1025 (reported by Schenkl; coni. Spengel) for *kath' hauto*.

7. *angeion* ('vessel') is the generic term for any given 'container' (*periekhon*).

8. This second claim is the second 'common notion' about place listed below at 111,7 (*ad* 211a2); cf. also 107,4-8 (*ad* 209b27-30).

9. 'In effect powers' (*hoion dunameis*: 103,6-7), or 'sort of powers'. See Algra (2), 196 n. 10 on the importance of this qualification (Themistius' addition) in connection with the general problem of natural place, which is denied the status of an efficient cause (105,11-12 below, at 209a22); see also White (2), 191-3, and Machamer for an extensive discussion. Themistius certainly does nothing here to encourage the view that place has an 'attractive influence' (Ross on 208b10), embodied in the Hardie and Gaye over-translation of *ekhei tina dunamin* (209a10-11) as 'exerts a certain influence'. His paraphrase of a related discussion of natural place in *Cael.* 4.3 is extant in a Hebrew version, edited with a Latin translation at *CAG* 5.4, but it would be unwise to try and exploit this material before it receives the new critical edition that it requires (see Zonta).

10. For *metastanti* (103,13) read *metastasi* (coni. Spengel). The plural is preferable given the antecedent *hêmas* (103,11).

11. 'So as' (*hôste*), Themistius' reading at 208b24, inherited from Alexander (see Simpl. *in Phys.* 526,16-18); *hôs* is transmitted by the Aristotelian manuscripts. See Ross on 208b24-5, and Mueller, 468-9.

12. Here Themistius is following Alexander in treating mathematical objects as abstractions. See Mueller, 467-70, who fails to note the present passage from Themistius, but cites another less directly relevant one from his *in An. Post.* 29,20-3 (which Mueller, 467 n. 9, mistakenly ascribes to the *Physics* paraphrase).

13. I place a comma after *eige estin* (103,26), clearly a parenthetical remark. Cf. 128,17 below for an almost identical qualification in another reference to the void.

14. *khaos* carries its literal sense of a gap or chasm at the start of the cosmogony; see Hesiod (late eighth century BC) *Theogony* 116.

15. See 102,6 (*ad* 208a29) above.

16. The verb *proüpoballesthai* translated 'be a prior condition' here is used at Themist. *in DA* 49,6.19.21 and 81,16 to describe the way that lower faculties of the soul are a supporting precondition for higher ones.

17. Literally 'what exists [in it]' (*ta onta*), which Themistius has substituted for the clearer Aristotelian phrase *ta en autôi* (209a2).

18. In 104,8 read *tôi* (MS **W**; Spengel) for the second *to*.

19. The Aristotelian text (209a2-4) is more abrupt: 'Nevertheless, if place exists, it is problematic what it is: is it a volume of a body, or some other nature? Given this, the first thing is to inquire into its genus.' Themistius incorporates the second half of this text into the opening of the first argument.

20. The verb is *aposemnunein* ('to puff up' or 'unduly dignify'). Themistius is probably echoing Plato *Theaetetus* 168D3 (a Platonic *hapax*, a lexical species for which this commentator always has a keen eye; see n. 348 below). There Protagoras is mocked for 'solemnly revering' his Man-Measure theory.

21. The direct speech attributed to these critics by Schenkl (it is not in Spengel's edition) should end at after *periakhthêsêi* (104,12) rather than after *sômati* (104,17); i.e. the rebuttal must come from the commentator himself.

22. Four of these arguments, nos (i), (ii), (v) and (vi), are addressed later at ch. 5, 121,21-122,5 (*ad* 212b22-9) below, while (iv), Zeno's paradox, is also handled at ch. 3, 110,22-111,3 (*ad* 210b22-7) below.

23. I punctuate *prôton – topon* at 104,12-13 as a question followed by an answer introduced by *ê*. I therefore change the indefinite *ti* to the interrogative form with the addition of an acute accent. Schenkl himself intended such a syntactical arrangement; see p. 251 of his Index Verborum, where he cites 104,13 as an example of the particle *ê* used 'post interrog.', in other words, to introduce the answer to an immediately preceding direct question, a usage, on which see Todd (6), 217. For Schenkl's insensitivity to it elsewhere, see below *ad* 107,9; 150,3; 152,1; 158,10-11.

24. Themistius' bridging paragraph at 104,9-12 correctly ensures that the claim that place is three-dimensional is part of the argument against the existence of place. In such a polemical context this statement cannot conflict with the obvious implication (never spelt out) that place is two-dimensional in the Aristotelian theory of it as the limit of the container. Sorabji at Urmson (1), 1 n. 1 is clearly right against Urmson *ibid.*, 22 n. 22.

25. The adjective *atopos* ('absurd'), means literally 'out of place' (*a-topos*); hence here (as at 107,26 and 111,2 below) it is used as a pun.

26. Chrysippus of Soli (280/76-208/4 BC) and Zeno of Citium (*c.* 334/3-262/1 BC), the third and first heads respectively of the Stoic school. Themist. 104,9-19 = *LS* 48F; 104,13-18 = *SVF* 2.468.

27. The addition of *en* ('in') at 104,18 Schenkl attributes to Diels; von Arnim (*SVF* 2, p. 152,12) also proposed it independently, two years after the appearance of Schenkl's edition.

28. On the topic of 'body going through body' in connection with the Stoic theory of total blending, and the criticism of it by Aristotelian commentators, based on texts such as the present one, see Todd (3), 29-88; the same study has a text and translation of Alexander's *De mixtione* (*On Mixture*), chs 5-9 of which deal with the Stoic theory in terms of the genus body, rather than with reference to Stoic qualitative physics. On bodily interpenetration see Lewis, Sorabji (3) chs 5 and 6, and White (1).

29. See above 102,20-103,4 (208b1-8).

30. At 104,25 *hosa* is a misprint for *hora* (correct in Spengel).

31. Simpl. *in Phys.* 531,4-30, suggests that since a point has no parts, then neither will the place of a point, and this will mean that there are two coincident points, which in turn means that there is only one point. Others (Hussey, 102; Bostock [2], 253) refer to an argument at Plato *Parm.* 138A3-7 which specifically argues that something partless cannot be in contact with anything in such a way that it is 'surrounded', or 'contained'.

32. This paragraph is Themistius' attempt to envisage conditions in which a place *might* be said to exist 'over and above' or 'in addition to' (*para*, 209a13) bodies. As he shows, this is incompatible with the general definition of place as a container, and so this option is driven back to the earlier position, that place is identical with, and thus indistinguishable from, its content.

33. At 104,33 I read *periekhôn* (Philop. *in Phys.* 511,25) rather than *periekhon*. Conceivably *topos* was construed as neuter in sense, but the implicit periphrasis <*esti*> *periekhôn* makes grammatical agreement preferable to the ambiguity produced by juxtaposing *mêden* and *periekhon*.

34. At 105,1 I read *isos* (MS **W**) and *ôn* (with MSS **WBL**), thereby replicating the text at Philop. *in Phys.* 511,25. The latter's paraphrase (511,25-6) is: 'but by being equal it is in just this respect a point, so that as a whole place will be a single point'. Themistius does not similarly restrict the argument to the point.

35. For *auto touto* as an adverbial phrase with isolating or restricting force see *LSJ*, *autos*, IV.1-3.

36. See n. 9 above on why the denial that place is 'what causes change' is significant for its status as a 'power' in the context of the definition of natural place.

37. Zeno of Elea (fifth century BC), a follower of Parmenides. See DK 29A24 on his paradox of place, which was also elaborated by Eudemus, fr. 78 Wehrli.

38. This reflects an Aristotelian text (209a23) in which the claim is that 'place will be somewhere'. Editions earlier than Ross had this clause as a question ('where will place be?'), a construal recently revived by Morison, 82 n. 5.

39. *hê* (105,13) is a misprint for *ho* (correct in Spengel).

40. This indirect reference to the void as a place unoccupied by body (cf. ch. 7 below, 124,29-30, *ad* 213b31-3) is Themistius' insertion.

41. Since *katekhesthai* is normally used in the passive voice of a place being occupied by a body, *panta katekhetai* (105,18) is a rare absolute use of the verb in the middle voice to refer to bodies 'in occupancy' of a place.

42. See 104,9-12 above.

43. At 105,22 read *legômen* for *legomen* with MS **B**.

44. Here the Aristotelian text (209a33) refers to the kind of place that is 'proprietary, in which [bodies exist] first' (*idios en hôi prôtôi*). This primary, or prior, sense of place is used more extensively in the paraphrase that follows than in the original.

45. This distinction is elsewhere formulated as that between the 'extended' (*en platei*) and 'exact' (*pros akribeian*) senses of place; see Sext. Emp. *Pyrrh. Hyp.* 3.37 and 119. It may have developed in Peripatetic exegesis. For similar language applied to a distinction between senses of 'the now' see 157,10 below.

46. *ouranos*, here the totality of the physical world, not the heavens. 'World' (Waterfield; Hardie and Gaye) is a possibility; 'heavens' (Hussey) an error. But 'cosmos' is best, given the English use of 'world' to mean planet Earth. Themistius himself substitutes *kosmos* for *ouranos* at 119,3 (*ad* 212b18-19).

47. Here this clearly refers to a conurbation rather than a city-state.

48. There is no reference in the Aristotelian text to place as 'proximately' (*prosekhôs*) limiting a body. The distinction between primary/prior and proximate was developed with reference to matter in the exegetical tradition, with prime matter a comprehensive concept distinguished from the proximate matter of which a substance was composed. See Accattino and Donini, 108, on Alex. *DA* 4,23-4. The claim that the relation of place to body does not work in this way anticipates the later argument of this chapter: that place is not matter or form.

49. This either means that nothing can physically intervene between a body and its place, or that, since primary and proximate are the only options, *tertium non datur*.

50. 106,4-26 is quoted as F96,2 by Gigon in his edition of the Aristotelian fragments, because of its similarity to material in later commentators assigned to the lost Aristotelian work 'On the Good' (*Peri tagathou*).

51. On the Aristotelian passage paraphrased here see Algra (2), 114-15.

52. This distinction would seem to be entirely Themistius' own conjecture; it has no corroboratory evidence in earlier literature. See Cherniss, 166 with n. 95. On the unwritten writings see Ross on 209b13-16. On the Platonic evidence (*Timaeus* 51A7-B1) for the theory of participation see Ross, 565-6.

53. A doxographical report ('Aetius' I.19.1, at *Dox. Gr.* 317) of Plato's theory of place has a similar reference to the metaphorical use of terminology.

54. At 107,1 read *toutôn einai ton topon* (MS **W**; Arist. 209b22), with *toutôn* taken as dependent on *hopoteronoun*.

55. At 102,20-103,4 (*ad* 208b1-8).

56. 'Affection' (*pathos*), *hexis* (transliterated), and 'power' (*dunamis*) are specific ways of characterising what elsewhere are described generically as 'incidental properties' (*sumbebêkota*), or just 'properties' (*huparkhonta*, literally 'things that belong'). They all meet the standard of inseparability from a substance.

57. *metaphorêtos* (107,7; 209b29); cf. below ch. 4, 118,23-119,2 (*ad* 212a14-21 and 28-30).

58. At 107,9 *ê* introduces the answer to the preceding question, not, as in Schenkl's punctuation (inherited from Spengel), a further question; the question mark after *epiphaneia* (107,10) must be replaced with a stop. For the answer to this particular question see 105,29-106,4 above (an expansive version of 209b1-2).

59. Themistius expands 209b31-2, possibly on the basis of *Phys.* 3.7, 207a35-b1 (sim. *Cael.* 4.5, 312a12-13), where the form/matter relationship is described in terms of containing and being contained.

Notes to pages 22-23

60. Themistius rephrases Arist. 209b32-3: 'And it is always believed that that which is somewhere is both something itself, and that there is something different outside it.' Ross, 567 clarifies this sentence much as Themistius does: 'if a thing is somewhere, there is something outside it, distinct from it; this, [Aristotle implies], is its place'.

61. On these 'eidetic', or 'form-producing', numbers see Themist. *in DA* 11,20-12,28 with Todd (6), 26-7 (and notes). The term 'eidetic' is not used in the present Aristotelian text.

62. Themistius turns the query at 209b34-5 as to 'why the forms and the numbers are *not* in a place' into the positive claim that they are in a place. Algra (2), 115-16 uses the negative emphasis in the Aristotelian text to argue that Aristotle, contrary to ancient and modern opinion (see especially Algra [2], 115 n. 91), did not think that the ideas themselves entered and left the receptacle. Clearly Themistius anticipated Simplicius (*in Phys.* 541,33-4), the only Greek commentator whom Algra cites for the standard view.

63. 'The great and the small', also known as 'the indefinite dyad' (*aoristos duas*), was in addition to 'the One', the Platonic first-principle of everything, including the forms. The doctrine was discussed by Aristotle in his work 'On the Good'. See Simpl. *in Phys.* 151,6-11 and 453,25-30 (= F93 Gigon).

64. Simpl. *in Phys.* 542,11 glosses this expression by quoting Plato *Timaeus* 51A7-B1 which refers to the receptacle 'participating (*metalambanon*) in a most perplexing way in what is intelligible' (tr. Zeyl). Urmson (1), 35 *ad loc.* mistranslates Simplicius by tacitly inserting a variant reading into the Platonic text (*ta porrôtata*, 'at the furthest remove', for *aporôtata*, 'most perplexingly').

65. The Aristotelian text is more elliptical. It just implies that there is some problem if water, as condensed air, occupies a place smaller than that which the air originally occupied. It is Themistius who applies this general query to the claim that place can be defined as matter or form.

66. In 108,1-2 I have removed the gratuitous brackets (not inherited from Spengel) around *phtheiretai – apodounai*, placed a colon after the second *topon* in 108,1 and a stop after *metabolais* in 108,2.

67. At 108,4 I read *autês* (MSS **WBL**; Spengel) (sc. *tês hulês*) for *autou* (MS **M**, Schenkl's favoured manuscript). The concluding clause of this sentence would otherwise have to mean implausibly 'and not separated from it [sc. the place]', instead of simply explicating the preceding clause.

68. See, for example, *GC* 320a1-5, and on 'prime matter' see n. 48 above, and cf. 106,17-18, a reference that like the present one is Themistius' own. Cf. also 137,23-138,18 below (*ad* 4.9, 217a21-33) on the basic doctrine of matter.

69. At 108,6 read *legômen* with MS **W** for *legomen*.

70. 'Like' or 'as' for *hôs* in this paragraph is perhaps elliptical, but less cumbersome than 'in the sense of', its full implication.

71. I follow Spengel's preference (reported by Schenkl) and add the definite article *ho* at 108,10 to parallel the way that species and genera are otherwise identified in this context.

72. For this sense and the next see the references at *LSJ* under *en* I.6. Since they are particularly remote from English idiom, some supplement is essential.

73. *Iliad* 17.514 and 20.435, and *Odyssey* 1.267 and 400 have the half-line 'it lies "in" (*en*) the knees of the gods' (our 'in the lap of the gods'). The knees were touched by the right hand in supplication; see, for example, *Iliad*

1.500-1. Themistius has perhaps quoted from memory in substituting 'Zeus' for 'gods'.

74. See Simpl. *in Phys.* 552,18-553,11 for such cases, of which the most obvious is that of being in time, extensively discussed in chs 10-14 below (especially at ch. 12, 154,12-16 [*ad* 221a9-13]; see the Greek English Index under *en khronôi*). Simpl. *in Phys.* 553,6-8 suggests that it was omitted because it had not yet been introduced into Aristotle's teaching.

75. 'By itself' will translate *auto* in this chapter. Themistius multiplies the Aristotelian usage at 210a25-6 where the problem posed is 'whether something just by itself (*kai auto*) can be in itself' (cf. also 210b13). Translators tend to omit 'by itself' in both loci; Themistius gives it full play, and redundantly uses *auto monon* ('all by itself') at 109,27 below.

76. See above ch. 2, 105,22-9 (*ad* 209a31-b1) on this general distinction.

77. This example, which is not in the Aristotelian text, is probably taken from Plato's *Symposium*, where the normally barefoot Socrates wore shoes to the tragic poet Agathon's party; see *Symp.* 174A3-4.

78. At 109,15 I place a comma rather than a stop after *leukon*, and read *hôste* for *hôs*, as suggested by D.J. Furley, rather than improving this sentence by emending *dunasthai* to *dunatai*, as Spengel proposed. Any pleonasm in *hôste dia touto* is tolerable; see 128,23 with n. 276 below.

79. *kath'hekaston* (109,18), Themistius' substitution for *epaktikôs* ('from examples', Waterfield).

80. At 108,7-20 (*ad* 210a15-24).

81. This is sense (ii) in the inventory above, i.e. 108,8 (*ad* 210a16-17).

82. At 109,25-6 the subject of the singular main verb *diapherei* can only be 'the essence' (*to ti ên einai*), and so the phrase conjoined with it, *kai hai phuseis kekhôrismenai* disrupts the grammar by creating a plural subject. It is probably an intruded gloss, designed to clarify the thought that something would be defined essentially as both content and container, and so have the 'different natures' referred to earlier at 109,16-17. But clearly *diapherei* has to lead into the separative genitive expressions at 109,26-7 and so *kai – kekhôrismenai* has to be deleted to remove the obstruction to its doing so.

83. That is, for X to be in itself, it has to be defined as X (i.e. have an essence), and then defined as X *qua* container, and X *qua* content.

84. Above 108,21-109,10 (*ad* 210a25-33).

85. This translates *monon auto* (109,27); cf. n. 75 above.

86. At 110,3 the subjunctive *enginêtai* (MS W) is clearly preferable to the indicative (*enginetai*) so that the purpose clause can be maintained at 110,1-3.

87. My translation follows Waterfield; literally the text has 'that in which' (the container), and '[what is] in this' (the content).

88. That is, some generic container (*angeion*), not the boat referred to at 110,10 above.

89. At 110,16 I have *faute de mieux* deleted *autôi houtô*, and eliminated *ho amphoreus* as almost certainly a gloss on *to angeion*.

90. See ch. 1 above, 105,12-14 (*ad* 209a23-5), and see Simpl. *in Phys.* 551,11-17 who regards the solution of Zeno's problem as the main purpose of the disambiguation of 'being in'.

91. See ch. 2 above, 105,29-106,4 (*ad* 209b1-2); cf. 107,10.

92. In sense (v) at 108,14-15 (*ad* 210a20-1) above health's presence in the body was identified as matter being in form. For Alexander's and Eudemus' objections on this point see Simpl. *in Phys.* 552,18-29.

93. At 111,2 I supply *hôs* before *en topôi* to make this statement parallel

the preceding claims about the way (or sense) in which health and temperature are in something, and to repeat 110,27 above; cf. 108,20 (*ad* 210a24).

94. In Zeno's problem; see 105,14 above.

95. There is no comment on 210b27-30 since it deals with the topic of matter and form being place, something already handled in ch. 2 above.

96. Literally 'hard to hunt down' (*dusthêratos*), a metaphor derived from Plato *Sophist* 218D3 and 261A5 where the sophist is said to be *dusthêreutos*.

97. These notions (*ennoiai*) are 'common' (*koinai*) in that they are considered indisputably accepted as fundamental (cf. *prôtai ennoiai*, 'primary notions', at 111,30 below). Their origin is in the Stoic argument from universal consent (*consensu omnium*). See further Todd (2) and Obbink.

98. At ch. 2, 107,4-8 (*ad* 209b27-30).

99. Here, and at 111,12 'content' translates *sôma* ('body').

100. See above ch. 1, 102,20-103,4 (*ad* 208b1-8).

101. (v) and (vi) omit some crucial material in the Aristotelian text at 211a4-6. In full that reads: 'that all bodies naturally move up or down and remain in their own places.' Themistius may have thought that it obvious that bodies moved to their natural places.

102. Aristotle (211a13-14) says that 'the main reason (*malista*) we think that the heavens are in place is that they are always (*aei*) in *kinêsis*'. Themistius has copied this sentence except for substituting *malista* for *aei* in its second clause. He may have wanted *kinêsis* there to be understood as a type of change, rather than movement *simpliciter*. He might have done better to have referred again to change in respect of place, as Philop. *in Phys.* 541,22 does. The present translation has had to insert that reference.

103. Aristotle has 'in increase and decrease' at 211a15-16, but *auxêsis* ('increase') can cover both these forms of organic change.

104. At 112,2 a stop rather than a comma is placed after *sunekhê*.

105. At 112,6 read *kath'auta* (MSS **BL**), for *kath'auto* (MSS **MW**; Spengel) at 112,6. Schenkl's apparatus criticus erroneously reports this information as applying to 112,5.

106. See common notion (iii) at 111,8-9 (*ad* 211a2) above.

107. At 112,2-4 above.

108. Aristotle at 211a31-2 says that '[the body] is in the extremity of the container as primary' (*en protôi esti tôi eskhatôi tou periekhontos*). Schenkl has Themistius saying (I think) that the body 'will be as in a place belonging to an extremity of the container' (*hôs en topôi estai tou eskhatou tou periekhontos*). But what it is in is the extremity, and so read *tôi eskhatôi* with a corrector of MS **W**, and delete *hôs*.

109. Literally 'are together' (112,17). Themistius substitutes 'together' (*hama*; cf. Arist. *Phys.* 6.1, 231a22-3) for Aristotle's 'in the same [place]' (*en tôi autôi*; 211a33-4).

110. Bracket *meizô – sômatôn* (112,21) and delete the colon preceding the clause; in this way it becomes an integral part of the sentence.

111. 'Hollow [or "inner"] surface' (*koilê epiphaneia*); i.e. the concave inner surface of a container such as the bulbous (externally convex) amphora used as an illustration later; see 116,18-27 below.

112. The phrase bracketed at 112,22 (*hê – keramiou*) reflects the guiding Aristotelian text (211b3-4), but its juxtaposition with 'extremities' is rather abrupt; contrast 118,26 for a similar explication introduced by *legô de*.

113. At 112,23 *sumballei* carries the relatively rare intransitive sense of 'fit together with'; see *LSJ* under *sumballein* I.4a.

114. In ch. 2, where it is discussed along with form; see 107,1-13, and 107,16-108,4.

115. There is an anakolouthon here. Literally: 'That it is also not form either – [well] what duped those who believed that place was form is by now obvious.'

116. That is, from the two universal affirmatives, *Form is a limit*, and *Place is a limit*, it does not follow that *Form is Place*, since 'limit' is ambiguous.

117. A stop instead of a comma after *arithmos* (113,7) creates the separate conclusion needed here. On number as a measure in counting see ch. 11 below, 148,25-9.

118. 113,7-114,2 = Usener, *Epicurea* no. 273 (194,13-195,9); 113,7-11 = *SVF* 2.506.

119. In the phrase *metaxu tês koilês epiphaneias*, here and at 113,21 (cf. 114,8), *metaxu* would be awkwardly translated 'between' rather than 'within', since in English it requires a plural complement, which it receives later ('extremities', 112,27 and 115,13; 'limits' 113,8; and 'hollows', 117,5) in descriptions of the inner space of a container that non-Aristotelians identify as place.

120. On Chrysippus see n. 26 and cf. *SVF* 2.503. 'Crowd', literally 'chorus' (*khoros*), Vettori's inspired emendation for *khronos* at 113,11. For *khoros* used with similar reference to a consensus within a philosophical group or school see, for example, Sext. Emp. *Math.* 7.60 and 9.1.

121. Epicurus of Samos (341-270 BC), founder of the Epicurean school.

122. In one half of the doxographical report at Aetius 1.18.1 (*Dox. Gr.*, 315) Plato is associated with the void in the sweeping generalisation 'Everyone from Thales to Plato rejected (*apegnôsan*) the void'.

123. At 113,29 read *ekrheuseie* (Usener) for *ekrhuseie*. Simpl. *in Phys.* 573,17-18, who is following Themistius here, mentions clepsydras as 'covered containers' that release water only when the air can get in; on their operation see Furley (1), 30 n. 25.

124. At 113,30 read *laboi* (coni. Spengel; Simpl. *in Phys.* 573,19) for *labêi* (MSS).

125. After his expansive version of Aristotle's first argument (211b14-19) against the concept of place as an extension, Themistius offers a series of related arguments (113,30-116,12) before picking up the text at 211b19 at 116,13. His rationale here (116,10-13) as later (see ch. 8; 132,5-7 below) is the obscurity of Aristotle's arguments.

126. Literally 'if the vessel fell together'; the verb is *sumpiptein*.

127. At 114,5 for *to dê* I read *tôi de*. The dative case with the infinitive is thus used in an explanatory sense.

128. Galen of Pergamum (AD 129 – c. 210-15), whose interests covered philosophy as well as medicine, was known to Themistius probably entirely from Alexander of Aphrodisias' commentaries. Alexander knew, or knew of, some of Galenic works (see Sharples [2], 1179), and wrote a treatise, preserved in part in the Arabic tradition, in which he attacked Galen's views on place and time (see Rashed, 318-31). Later discussions of the present argument on place (Simpl. *in Phys.* 573,19-29, and Philop. *in Phys.* 576,12-22) do not mention Alexander, but Themistius was not necessarily their exclusive source; see Nutton, 48 n. 67. On Galen and Aristotle, and the present Themistian text, see Moraux (2), 729-30 with n. 170, Gottschalk, 1166-71 with 1168 and n. 418, and Richard Sorabji in the Preface to this volume, p. viii.

129. Philop. *In Phys.* 576,22-577,9 (after quoting 114,7-12 at 576,16-21)

claims that Themistius was unfair to Galen. In his view, Galen was a precursor of his own theory that place was an extension that is never empty of body, and his hypothesis was designed to establish this, and not the existence of a 'separate extension'. In fact, Galen was probably just raising objections to Aristotle, without any systematically constructive intent; see Gottschalk (preceding note).

130. At 114,12 I read *holôs te ouden* (*oute* MSS) *dunaton*. The sense is obvious.

131. At 114,14 read *tou legein* for *to legein*.

132. At 114,14 *ei tis hupothoito* (equivalent to *kath'hupothesin*) should be enclosed by commas as a parenthesis.

133. cf. Cleomedes, *Caelestia* 2.1.401-3 for an analogous conceit in a passage about the power of the Sun: that without it everything will be eliminated.

134. Philoponus *in Phys.* 575,27-576,12 offers a response to this claim.

135. The void can be crudely defined, as it is implicitly here, as place without body; see ch. 7, 124,29-30 (*ad* 213b31-4).

136. It is not entirely certain, but this question and the ensuing dialogue does not seem to be between Themistius and Galen, but to be directed to an imaginary interlocutor.

137. To say that a body could not be fully in contact with a *flat* surface such as a wall would be to claim that there was always space intervening between bodies, and hence to deny that matter was continuous; that would mean asserting the existence of the void.

138. The logic of the question at 114,21-3 is that of a leading question anticipating the answer 'No', notwithstanding the fact that its main verb is negated by *ou* rather than *mê*. Its gist is that you ought to believe that all bodies, whatever their shapes, are fully in contact with an adjacent body.

139. Literally, *monon kath'heauto*, 'alone in respect of itself (or *per se*)'.

140. This is probably an allusion to Arist. *DA* 3.10, 433b14-19, where the role of desire (*orexis*) (Themistius uses the Stoic term *hormê*, 'impulse') is defined in relation to the other elements involved in a rational animal's movement.

141. See 111,6-15 above. Themistius here uses items (i) and (iv) from that passage.

142. This is a reference to Galen, as analysed at 114,17-21 above.

143. This is implied by 115,2-10, where different bodies enter an extension of the same quantity that is erroneously believed to be its place, rather than the extension proprietary to each body.

144. cf. 113,26-9 above for the extension as associated with body and distinguished from place.

145. I have placed a colon instead of a comma after *autôn* (115,26) so that what follows can be the direct question that Schenkl makes it, rather than the indirect question it is on his existing punctuation. The half asyndeton at *hotan* etc. after the proleptic *touto* (115,25) supports this rearrangement.

146. At 115,28 I read *en tôi tou topou* (sc. *diastêmati*) with MS L rather than Schenkl's *en tôi topôi*. The logic of this sentence requires a reference to the extension of the place.

147. Here *metron* has to identify the dimensional space that the extension will occupy; for the sense see *LSJ*, *metron* 3.

148. The elliptical nature of this sentence suggests that the entailment I have explicated was perhaps displayed to the students. The first clause is implicitly a conditional, since the proponents of the extension theory would not generally accept that bodies coincide. I have tried to reflect this in the translation. Syntactically *ê* (116,1) introduces the answer to the question at 115,33-5, but does so with a hypothesis; i.e., if they accept the proposed entailment, we can deny the consequent (or, put otherwise, we can refute them by *modus tollens*).

149. That is because a surface is two-dimensional, even though this implication is not spelt out by Aristotle or Themistius; see n. 24 above.

150. At 116,11 I read *asaphesteron <ti> men ekhei*. Cf. Cleomedes, *Caelestia* 1.7.50 for identical language.

151. This is not a reference to the start of the paraphrase of the *Physics*. The goal of 'uncovering' (*ekkaluptein*) Aristotle's obscurities is part of the general programme for paraphrasing that Themistius enunciates elsewhere (*in An. Post.* 1,16-2,4); see Todd (6), 3-4.

152. 116,12-117,3 is quoted with some rephrasing at Philop. *in Phys.* 550,9-551,20.

153. I read *en tautôi* ('in the same [place]') at 116,13 with Philop. *in Phys.* 550,10 in order to reflect the Aristotelian text at 211b20. Themistius' MSS give *en heautôi* ('in itself'), which is implausible. With this change a comma needs to be placed after *menein*. See Ross on 211b19-21 on the text and syntax here.

154. At 116,14 I have deleted *ê aeros* ('or air') as a gratuitous addition originating probably with an annotator; the rest of the sentence requires reference exclusively to water.

155. At 116,20 read *hou diastêmatos* with MS **W** and some of the manuscripts of Philoponus *ad in Phys.* 550,17.

156. This is the extension that the total volume of water has independently of the container through the amphora's extension doing double duty. The text at 116,25 needs emending. It has to say that extension (vii) is that of a body, which can only be the water in the amphora. The text at 116,26 has an unintelligible sequence after 'the [extension] of the amphora', namely, *hôs tou hudatos sômatos*. But we need *hôs sômatos* here as in the next clause, and can delete *tou hudatos* as originally a gloss on *sômatos*; in fact, in the translation it is reinstated as an explanatory supplement.

157. The difference between (viii) and (vi) is that in (viii) the part is considered as an independent body, with its own extension, definitionally distinct from that of the amphora, whereas in (vi) a part of the water occupied a place that was definitionally identical with the inner surface of the amphora. Case (viii) not surprisingly elicits criticism in the digression in the next paragraph.

158. As Sorabji (3), 76-7 notes, Themistius may not be denying that the extensions can coincide, but 'may rather be thinking that since extensions *could* coincide, so (absurdly) would places, if they were extensions'.

159. The argument is 'from outside' (*exôthen*), i.e. not from Aristotle, whose critique at 211b25-9 ends just before this parenthesis. Simpl. *in Phys.* 576,21-3 repeats this argument, and also says that it is *exôthen*, which I follow Urmson (3), 73 in translating 'a digression'; the sense is not recorded in *LSJ*.

160. This elliptical paraphrase omits 211b31-4 (an illuminating reference to the role of matter in qualitative change [*alloiôsis*]), and 211b36-212a2 (a summary), while it introduces a reference to the relation between bodies and their places that is not in the Aristotelian text.

161. For the arguments against these identifications see ch. 2 above, 107,1-13 and 107,16-28, and in the present chapter at 112,28-113,8.

162. See 112,25-8 (*ad* 211b5-10) above.

Notes to pages 31-33 85

163. This final clause (*kath'ho sunaptei tôi periekhomenôi sômati*, 118,8-9) is taken to be part of the Aristotelian text (212a6-6a Ross), although it is missing from all the Greek manuscripts; it is found in all the Greek commentators and in the Arabic-Latin translation. See Ross *ad loc.*

164. This remark, which has no basis in the Aristotelian text, is out of place here, except as a proleptic and elliptical aside. It mentions a distinction that Simplicius (*in Phys.* 589,5-8) quotes Alexander as having appropriately drawn in connection with *Phys.* 4.5, 212a31-b3, where the circular motion of the heavens is under discussion, and the revolution of the parts is contrasted with the immobility of the whole (see Themist. 119,23-4 below), a situation that is indeed distinct from locomotion.

165. That is the place that fully coincides with its content; see 105,26-7 and 105,29-106,4 (*ad* 209b1-2).

166. At 113,8-30 above (*ad* 211b14-19).

167. Aristotle at 212a10-12 is rather clearer. He explains the mistake as follows: 'it seems possible that there is an extension between [the limits of the container] that is something distinct from the magnitudes of the moving bodies'. In other words, two magnitudes of equal volume occupy at different times the volume of the container, which is true, but that does not make the container's volume their place (at least for him).

168. The use of *kleptesthai* ('be disguised') fits with the earlier claim that those who believe that place is an extension are 'duped' or 'tricked' (see 112,30 and 113,30).

169. This last comment is Themistius' further rationalisation. It is unclear what sort of rapid replacement and what sort of vessel he envisages. The best example might be the way that the apparently incorporeal air referred to in the next argument replaces a solider substance ejected from a vessel.

170. 212a21-8 are incorporated in the opening paragraph of the paraphrase of ch. 5 at 119,3-12 below.

171. This parenthesis, a condensed version of 212a29-30, is all that Themistius takes from 212a28-30.

172. Schenkl (on 119,1) claims that Themistius passed over 212a21-6. But he has in fact used 212a21-31 as the basis for reorienting the next text (212a31ff.) to the general question of how the heavens can be moved when they have no containing limit. Schenkl was thus justified in moving this paraphrase to ch. 5. This passage, it should be noted, invokes four of the six 'common notions' identified above at ch. 4, 210b34-211a6 (= Themist. 111,6-14), specifically (v), (vi), (i) and (iii) in our translation, in that order.

173. 'Whole [cosmos]' translates *ta hola*, a collective plural used in this sense in numerous cosmological texts (see Todd [5]), and employed here as a semantic variant in this explicit gloss.

174. The heavens move, but still retain the same place in their totality. They thus 'appear' to be unmoved when considered under that description. See 120,17-19 below.

175. If the centre of the earth is a point, then it cannot be physically stable, since it will be partless. But 'what is beside the centre' (<*to*> *para to meson*; I have supplied *to* at 119,8) must refer to whatever quantity of matter is needed to create a stable cosmocentric mass.

176. Themistius has directed this text to the subject of the heavens by taking *to pan* at 212a34 to refer to 'the all' (the whole cosmos), and can thus focus on the issue of the movement of the heavens raised in the later part of this text, and ignore the comparison with water at 212a32-4.

177. The sentence at 119,21 calls for matching parentheses (cf. 119,14-16 above). Thus I have removed the stop at line 20 after *kineitai*, placed brackets around *hautê – topos* (20-1) and followed it with a comma. This arrangement matches it with the bracketed clause *hautê – methistamenôn* (19-20).

178. Eudemus of Rhodes (second half of the fourth century BC), a member of Aristotle's school, and author of an extensive commentary on the *Physics*. Wehrli at Eudemus fr. 80 quotes Simpl. *in Phys.* 595,3-15, of which lines 9-15 correspond in all but minor details to the present Themist. 119,27-120,3. For an alternative translation and discussion of the Simplician evidence see Algra (2), 256. On Eudemus on place see Sharples (4).

179. I take *holôs* at 119,28 in this logical sense, as a reference back to the general theory of place as an external limit. Urmson (1), 94 takes it in a physical sense: 'they are not in place as a whole'.

180. My translation of Eudemus differs from that at Urmson (1992), 94. I note especially his translation of the final clause (*pollakhôs de to pou*, 120,3): ' "somewhere" is ambiguous', where he construes *to* as equivalent to quotation marks for the indefinite term, while I take it as referring to category Where (Arist. *Cat.* 2a1-2, and *Metaph.* 5.7, 1017a26). Probably *to pou* here reflects Arist. 212b14, where it is best taken as 'for a thing to be somewhere' (Waterfield). Thus 'being somewhere' has been shown to have one sense when used of the parts of a whole, and will shortly be shown to have another sense when the heavens are said to be in place incidentally.

181. The reference back is also in Aristotle, and Ross takes it to be to ch. 4, 211a17-b1, where the distinction employed here is at least implied in somewhat different language in a similar discussion of place as applicable to something that is discontinuous and that can therefore have a limit.

182. This sentence is Themistius' elaboration of 212b10-11: 'to the extent that the heavens move, its parts also have a place; for one [part] is next to another'. The contrast of circular celestial motion (*peripheresthai*; 'being carried round') with locomotion (*pheresthai*, 'being carried along') is Themistius' addition.

183. Simpl. *in Phys.* 592,11-593,6 has a quizzical discussion of this claim; he refers to Themistius (see n. 188 below), but not to any pre-Themistian commentator. Themistius is undoubtedly referring to Alexander.

184. At 212b11-12.

185. This is a reference to ch. 3 above, 109,10-18 (*ad* 210a33-b8). The phrase 'in respect of a part' (*kata meros*) is Themistian shorthand for Aristotle's statement at 210b1-2 that certain things are said to be in themselves when they are described 'in respect of things that are parts' (*kata tauta ... merê onta*).

186. See Themistius, *in DA* 18,30-7 for another example of this kind of usage.

187. See King, 80 n. 1 on the later history of the problem to which this is Themistius' solution, which was accepted most notably by Aquinas.

188. Themistius' position here is criticised by Simpl. *in Phys.* 592,22-593,6 (cf. 590,27-32), who quotes Themist. 121,2-4 at 592,23-4. He recognises that Themistius has taken Aristotle's 'in a way' (*pôs*), which is used at 212b12-23 to qualify the expression 'all the parts are in place', so that it specifically qualifies 'in a place'. This allows Themistius to argue that the outermost sphere is 'in a way contained' by the sphere on its inner side. I take it that the force of *hoion* before *periekhetai pôs* at 121,3 is designed to draw special attention to *hôs* as Aristotle's term; i.e. 'as [Aristotle says]'. See also the Preface to this volume, p. ix.

189. At 121,5-6 I make this question a parenthesis in the ongoing sen-

tence; i.e. I delete the colon after *dunamei*, and place brackets around *pôs – holês* and follow it with a comma.

190. In explicating the idea of something over and above the All (212b15-16) Themistius can presuppose Aristotle's discussion of the infinite in *Physics* 3.4-5, in particular the argument (3.4, 203b22-30; cf. Themist. *in Phys.* 81,30-82,18) that outside the heavens there is an infinite void (cf. Arist. 203b27-9) that must be occupied by infinite body.

191. If an extracosmic body is in place, then it must be limited by another *body*, on the Aristotelian theory, and so on *ad infinitum*. For something close to this argument, used against the Epicurean theory of extra-cosmic body, see Cleomedes, *Caelestia* 1.1.132-8; he, however, envisages body being bounded by void, and void by body, in an infinite series, since, as a Stoic, he can accept the existence of infinite, but unoccupied, extracosmic void.

192. He is probably referring to the arguments in *Phys.* 3.5, rather than or additionally those at *Cael.* 1.7, since the Themistian *ordo expositionis* would seem not yet to have reached the *de Caelo* (see n. 237 below).

193. The definite article *to* is supplied before *pan* at 121,15, as at Arist. 212b18.

194. While 'cosmos' again (cf. n. 46 above) translates *ouranos*, it later (121,20) reverts to its restricted sense, i.e. the heavens.

195. The verb at 121,16, *periblepein* in the active voice means 'look at'; but one of the senses assigned its middle voice by *LSJ*, 'seek after' or 'look for', is required here.

196. The problems solved here are four of the six arguments against the existence of place listed in ch. 1, 209a2-30 (Themist. 104,12-150,20). Using the numbers supplied in the translation of both passages, (i) here responds to (vi) above (105,14-20) (ii) to (ii) (as at 104,25-30), (iii) to (i) (104,12-17), and (iv) to (v) (105,12-14).

197. Aristotle here (212b29) refers only to *to kinêton sôma*, and translators differ on whether to render this 'moveable body' (Hardie and Gaye; Waterfield), or 'changeable body' (Hussey). Themistius' addition of *kata phoran* ('in respect of motion') ensures that locomotion is represented as a species of change, not its only manifestation, and not the only kind of change relevant to being in a place.

198. At 122,10 clearly *tôi*, proposed by Schenkl in his apparatus, should be read for *to*.

199. That is, in relation to the body that forms their container.

200. This reference is also in the Aristotelian text (213a4-5). Commentators take it to be primarily to *GC* 1.3 (cf. Simpl. *in Phys.* 599,3-4), and *Cael.* 4.3.

201. This is a very elliptical comment, perhaps a note to be expanded in oral delivery. The Aristotelian text refers to the potentiality of water (to become earth or air) as analogous to the relation between part and whole, and contrasts this with the relation between two organically fused bodies, which are indistinguishable. Themistius puzzlingly turns this fundamental contrast into an analogy between bodies that are in contact and those that are fused.

202. This (unsolved) problem is added presumably to bring the discussion back to the central theme of ch. 4, the rejection of the theory of place as extension.

203. I have deleted *touto* at 122,27; it is improperly positioned, and would at most add a gratuitously ostensive force to its phrase.

204. On the procedure for examining place see 102,2-4 (*ad* 208a27-9) above.

205. This is the language of the conclusion to the chapter (213b27-9), later omitted (see n. 221 below), rather than that of the Aristotelian text at 213a20-1.

206. These 'standard beliefs' (*koinai doxai*) are those of its proponents; see 124,29-125,26 (*ad* 213b31-214a16) below.

207. Literally 'at the doors' (*kata thuras*, 123,1 = 213b2). This probably reflects the proverb at Arist. *Metaph.* 2.1, 993b5, 'Who could miss the doors?', i.e. miss the plain truth.

208. They are hand-held devices to draw and retain, and the water is kept in them by holding the thumb over an aperture. For a detailed description see Furley (1), 30 n. 25, and cf. 133,22-5 below for their use in an experiment designed to disprove the existence of the void.

209. This conflates the disjunction in the Aristotelian text at 213a32-4 where it is said that the extension distinct from bodies should be shown to be 'neither separable, nor as actually existing, in that it divides the totality of body so that it is not continuous'. That is, Themistius eliminates the potential ('separable') void, and refers only to one that is 'actually separate'. As Thorp, 151 n. 10 observes, the separable void would not cause discontinuity in body, even though translators of the Aristotelian text proceed as though it can by not punctuating Aristotle's text so that this type of void is clearly contrasted with the actual void. Themistius eliminates the whole distinction.

210. Here (123,15, also 123,20) *ouranos* is used in the sense of the whole cosmos, not the heavens; the reference to the whole of body (*to pan sôma*, at 123,14 echoing 213b1-2; cf. Arist. *Cael.* 278b21-4 and Themist. *in Cael.* 53,26-8) makes this obvious.

211. Literally, 'sown alongside' (*paresparthai*, 123,16). The verb is not in Aristotle, and may originate with the Epicureans; see, for example, Epicurus, *Hdt.* 63 where it is used of the soul's relation to the body. It conveys the idea of wide diffusion in an interstitial form. It was perhaps employed by Strato of Lampsacus in connection with an intracosmic micro-void.

212. Democritus of Abdera (*c.* 460-350? BC), and Leucippus (fl. *c.* 435 BC), were the founders of Atomism. At DK 67A19 the argument beginning at *Phys.* 213a27 and including most of the rest of this chapter is cited as relevant to the Atomist theory of the void.

213. This extracosmic void, which Aristotle describes here (213b1-2) as 'something outside the whole of body', Themistius frequently characterises (without Aristotelian precedent) as *athroos*, a term applied to a large mass, or collectivity, and translated hereafter as 'gross' or '*en masse*'. Later (4.9, 217b20-1) Aristotle calls such a void 'separate unqualifiedly' (*khôriston haplôs*).

214. 123,15-22 = Usener, *Epicurea* no. 274 (195,10-18); 123,20-2 = *SVF* 1.94. For the Stoic argument for an infinite void existing outside a finite continuous cosmos see Cleomedes, *Caelestia* 1.1.20-149.

215. The parenthesis *kinêsis – phthisis* at 123,25 reflects Ross' punctuation for the related phrase at Arist. 213b5.

216. The sentence at 123,32-3 has to be divided into a rhetorical question ending at *allo*, and an inference introduced by *kai houtôs*. Schenkl failed to punctuate the whole sentence as a question, despite its beginning with an interrogative.

217. Alexander turned this argument for the void against the Stoics, whom

Notes to pages 37-38 89

he saw as claiming that a body went through a body; see Todd (3), 80-1 and 86-8. See especially Alex. *ap.* Simpl. *in Phys.* 530,19-24.

218. Any given volume can be divided up into smaller units that will be equal to the volume of any smaller body. These units will be produced by being divided 'in successively different ways' depending on what the size of the smaller volume is. The result will be a set of equals (i.e. all equal to the smaller volume); these are numerous, and also collectively unequal *vis à vis* the volume of the smaller body. The translation 'are created by division' comes from construing *diairoumena ginetai* as a periphrastic present with emphasis on the result of the process.

219. At 124,2 place a stop after *ginetai*, and start a new sentence with *hôste*; for the latter in initial position to 'mark a strong conclusion' see *LSJ* s.v., II.2, and cf. n. 313 on 133,1 below. See also 160,9 below.

220. See Melissus (the Samian Eleatic; fifth century BC) at DK 30B7, scts 7-10.

221. It would be rash to assume that Themistius' expository inversion of 213b14-18 and 18-20 necessarily reflects his manuscript authority.

222. Here the term is used in the sense of organic growth.

223. See Alexander, *Mixt.* ch. 16 for an extensive defence of Aristotle's theory of organic growth as not requiring a body to go through a body; for the complementary argument that the phenomenon need not be explained by interstitial void spaces see Todd (1) on Alex. *Quaest.* 2.12.

224. See ch. 9 below for a detailed analysis of the void as the purported explanation of bodies becoming more, as well as less, dense.

225. Imagine a large cask ostensibly full of wine, and a wineskin full of wine being inserted into it, without any resulting overflow. This allegedly happens because the wineskin has void spaces within it that can be occupied when its contents are submerged and subject to pressure. This example differs from that at the pseudo-Aristotelian *Problems* 25.8, 938b14-24, where air (rather than the void) is squeezed out of the wineskins that are inserted into a cask, thereby creating space for additional wine to be poured directly into the cask.

226. That is, the vessel is supposed to be full of ash that becomes totally saturated with a volume of water equivalent to the volume of the vessel; i.e. the water does not overflow, and none of the ash is displaced. On the objection to this evidence see n. 266.

227. 213b27-9, a reference back to 213a20-1 (= 122,31-2 above), is omitted; see n. 205 above.

228. Literally, 'unless the void fell in alongside' (*parempiptein* is the verb). This is a non-metaphorical equivalent to *paraspeiresthai* ('be disseminated') (see n. 211 above).

229. This opening sentence is Themistius' attempt to put the Aristotelian report of the Pythagorean position into the context of a general argument for the void.

230. Themistius depends on Aristotle's report here; Ross on 213b23-4 and 24-7 should be consulted.

231. See 122,32-123,11 above. 'In relevant terms' (*pros epos*); see Plato, *Philebus* 18D6.

232. I read *legômen* (124,24) (hortatory subjunctive) with Vettori for *legomen*.

233. *anupostatos*, i.e. is without *hupostasis*, 'subsistence', the condition of real existence; in other words, it has no reference.

234. This was a position that Themistius inserted into his paraphrase of 213a12-22 at 122,28-9 above.

235. All void is unoccupied by body; all body is an object of touch (= has heaviness or lightness); therefore, all void is unoccupied by heaviness or lightness.

236. That is, they would propose 'void = df. a place in which no body exists'.

237. This aside, which is not in the Aristotelian text, might be interpreted as confirmation that Themistius' students, as we might expect, would study *de Caelo* after the *Physics*.

238. At 125,17 I read *tode* <*ti*> , in the light of Aristotle's text at 214a12.

239. cf. above ch. 2, 107,12 (*ad* 209b21-4).

240. *touto de autôn pariasin* (125,25) is difficult to construe. *parienai* with the genitive case should mean to release something, so that more intelligible Greek might be *toutou de autoi pariasin*, i.e. 'they themselves release this', in the sense that they do not get a firm conceptual grip on the void that they are positing. I have translated the general sense required.

241. This is a reference to the main arguments in ch. 4 above; note the earlier references to the void as equivalent to place *qua* extension at 113,10, 114,20-1, and 115,18-19.

242. At 122,27-9 above.

243. See n. 328 below for a discussion of the translation of terms for relative density, and a defence of the use of 'compact' for *pakhus*, the adjective employed here, and *puknos*.

244. See ch. 6 above, 123,7-11 (*ad* 213a24-7).

245. i.e. a 'gross' void, outside the set of all bodies.

246. *ek tôn autôn hormômenoi* (126,10); for the same verb used to describe the process of advancing into argument from a basis see Cleomedes, *Caelestia* 1.5.105; 1.6.103; 2.1.275; 2.2.6. Cf. also the use of *hormêtêrion* at Themist. *in DA* 1,26.

247. See 123,23-124,4 above (*ad* 213b3-14).

248. At 126,10-13 the flow of the reasoning is clearer if the colon before *apo gar* etc. (126,10-11) is removed, and *apo – sômatôn* bracketed, and followed by a comma.

249. Schenkl should have printed <*kai*> *tauton* at 126,14. That is, he seems to have emended *kat' auton* in MS **M** with reference to *tauton* in MSS **WBL**. But **M**'s reading is surely itself an emendation of *tauton*, and therefore its *kat'* has no independent value, and Schenkl's *kai* cannot be a further emendation of it, but must be identified independently.

250. See the elaboration of the paradox of bodily interpenetration at ch. 6 above, 123,29-124,2 (*ad* 213b7-12); cf. also ch. 1, 104,12-22 (*ad* 209a4-7).

251. With Spengel I read *exetazômen* (126,17) for *exetazomen*.

252. In this analysis Themistius, like Aristotle, is assuming that the proponents of the void operate with a distinction between types of change. However, as the paraphrase of ch. 9, 216b22-5 unwittingly reveals (see 135,13-17 below with n. 329), it makes more sense to assume that the presence of void in bodies reduces all physical change to change of place.

253. See ch. 6 above, 124,3-4 (*ad* 212b12-14) for Melissus' argument that if there is no void, then there is no movement at all in the universe.

254. The sentence at 126,17-22 emerges more clearly if we remove the colon after *legousin* at 126,18, bracket the clause *ou gar – elathen* (126,18-19), and follow it with by a comma rather than a stop. A corresponding parenthesis can then be established by bracketing *dunatai – kekhôrismenou* (126,20-2), and following it with a stop.

Notes to pages 40-41 91

255. See 124,9-14 above (*ad* 213b14-18), where this is the third argument.

256. For *expurênizontos* at 126,29 read *ekpurênizomenou* with MS **M**; the present passive is required for coordination with the use of this verb in the next clause.

257. 'Extension' here translates *diastêma* (127,1) which carries a special sense of the *additional* extension added to a body rather than its usual sense of a static three-dimensional volume.

258. See above 124,4-9 (*ad* 213b18-20), where this is the second argument in that chapter.

259. i.e. in the sense of organic growth; cf. the reference to nutriment at ch. 6, 124,6 (*ad* 213b19-20).

260. This widespread saying (equivalent to the British 'hoist by his own petard') seems to have originated in Aeschylus' lost tragedy *The Myrmidons*; see Nauck, 45-6 and Radt, 252-5 (the latter includes the Themistian text). Philop. *in Phys.* 577,3 used it in challenging Themistius' arguments against the void.

261. At 127,13 Vettori's emendation of *de* to *dê* after *hora* is essential for syntactical continuity. Cf. 116,21 where *hora dê* is found in Philop. *in Phys.* 550,18 as a variant for Themistius' *hora de*. For *hora dê* as a Platonic idiom see Denniston, 217.

262. The verb is *kataballein*, famously used by Protagoras in the title of the work (*hoi kataballontes*) in which he tried to 'floor' everyone with his Man-as-Measure proposition. See Sext. Emp. *Math.* 7.60 (= DK 80B1).

263. There is no comment on the concluding sentence 214b10-11.

264. 124,14-15 (*ad* 213b21-2 above), which there, as here, is fourth in the sequence.

265. A *kylix* is introduced here because it is a wide and relatively shallow basin in which the absorption of water by ash could be easily observed.

266. A fair point, if the argument is that absorption is possible only into void spaces, with no account being taken of the inherent properties of the ash. Such an argument is in fact a sitting duck: for unless the *kylix* is completely empty, no type of matter can ever absorb water greater in volume than the unoccupied portion of the vessel (just as if increase is by entry into void spaces, it can occur throughout a body only if the body is a complete void; cf. 127,15-16).

267. See n. 226 above where this statement is elicited from the elliptical description of the argument given.

268. The pseudo-Aristotelian *Problems* 938b24-32 accepts that a container of ash can absorb an equivalent volume of water, and explains the phenomenon as a process of slow saturation, due presumably to the nature of porous matter.

269. That is, the four problems raised in ch. 7 (change, varying density, increase, and absorption of fluid by porous matter) can be addressed in Aristotelian terms, without any polemic against the void, as ch. 9 below shows for the case of varying density, or *GC* 1.5 for the case of increase.

270. At 128,5 I read <*ou*> *mallon*, a standard locution (cf. 122,23-4 above; 128,27-8 and 130,3 below). The received text makes this final clause illogically say that 'the void is the cause of stability more than that of motion'.

271. Themistius does not reproduce Aristotle's phrase '*kinêsis* in respect of place' (214b17), and just uses *kinêsis* (translated 'movement' here) in the sense of locomotion.

272. See ch. 5 above, 122,5-14 *ad* 212b29-213a6.

273. At 128,9 *ephelkesthai* cannot mean 'drawn' or 'attracted'; it has to cover an unquestioning acceptance of the position identified. *LSJ* III.5 offers 'claim for oneself, assume' with reference to Plato *Gorgias* 465D5. Themistius' usage is close to that.

274. See ch. 4 above, 113,8-10.

275. See ch. 4 above, 116,12-117,24 (*ad* 211b19-29).

276. At 128,23 the deliberative subjunctive *legômen* with *an* is unorthodox. It would probably be best to read *au* and treat *au palin* as a case of pleonasm (directly paralleled at Themist. *in DA* 44,31 and 49,1), and found in other forms in the present work at 109,15 (*hôste dia touto*; see n. 78), 111,11 (*eti pros toutois*), and 132,14 (*men oun houtô*).

277. Aristotle had only said 'because of the uniformity' (*dia to homoion*), to which Themistius has added 'of its container' (*tou periekhontos*), i.e. what surrounds it. He does not, however, explain what this uniformity involves. Ancient and modern commentators alike refer to Plato's argument (*Phaedo* 109A; *Timaeus* 62D) for the earth's stability being the result of an 'equilibrium' existing between it and its medium: e.g., it is said to be 'positioned in the middle of something that is uniform' (*homoiou tinos en mesoi*; *Phaedo* 109A). Ross on 214b31-2 simply follows Simpl. *in Phys.* 666,23-6.

278. At 129,1 I read *mê têi kata phusin antikeimenê*; i.e. the forced motion is itself opposed to natural motion. Schenkl's dative case, *antikeimenêi*, would mean that motion was 'forced upon opposed natural motion', whereas Spengel's conversion of the clause to a genitive absolute is grammatically possible, but retains the same idea of natural motion being in opposition.

279. At 129,8-9 the clause *to – Dêmokritos* is best bracketed as a parenthetical remark, with the colon preceding after *kenou* (129,8) deleted.

280. A comma after *tautês* (129,11) is needed in order to isolate this clause.

281. One non-natural movement, that is, gives rise to another, as we see in the description of ballistics at 129,19-27.

282. 'Is gone' (*oikhetai*, 129,12), as in 'gone for a Burton' (UK) or 'out the window' (US) (cf. *LSJ* II.1b, or II.2), a colloquial equivalent for *anaireisthai* ('be eliminated').

283. At 128,24-5 (*ad* 214b28-30).

284. This sentence needs to be punctuated by a comma at 129,15 after *phusin*, so that its contrasting clauses can be more effectively juxtaposed.

285. On throwing see also Arist. *Phys.* 8.10, 266b27-267a12, with Themist. *in Phys.* 234,12-235,29. Sambursky, 71-2 used some of this latter paraphrase to try and characterise Themistius as a forerunner of the impetus theory of imparted motion; see the next note.

286. *epôthein* (129,22) is a vigorous forward thrust (as opposed a nudge forward, conveyed by *proôthein* at 135,18 below). The sense of the compound is retained in the simple forms at 129,23 and 26 by a well-established syntactical principle; see Renehan, 11-27.

287. It is important to translate *sunôthoumenon* as middle rather than passive. The air at the front of the missile is agitated enough to assist the trajectory. The difficulty of describing this process is evident from the qualifying *hoion* ('in effect'), which is repeated when at 129,28 'dragging' is used to describe this process.

288. This translates *endotheisa kinêsis* (129,25), a phrase that, as Wolff, 104 notes, should not be confused, as it has been, with the impetus theory of dynamics; here it refers only to a movement being passed on to the medium, and not to the transmission of any force.

289. Themistius' elaboration of the brief Aristotelian account of dynamics is undoubtedly derived from Alexander, and is found in a more developed

form in Simplicius *in Phys.* 668,24-669,15. It is a footnote in a larger debate that involved Philoponus' impetus theory; see Sorabji (3), 144-5 and 228-9, Lautner, 231 n. 92 and, above all, Wolff.

290. At 129,27 a stop after *kenôi* is too emphatic. I have replaced it with a comma, and refer to 108,23 where *alla* also introduces a coordinated clause after a comma.

291. Thrusting forward by air is the mechanism in this first explanation of ballistics (129,20-2 above); dragging (*helkein*) enters the picture in the second (129,23-7) in the form of 'co-thrusting' (*sunôtheisthai*) by the air, which is in effect a dragging forward of the missile.

292. The question mark (inherited from Spengel, 294,2) after *pou* in 130,3 must be replaced with a stop; the sentence contains an indirect, not a direct, question.

293. At 130,4 I have ignored *hôs te*, and translated this question as Themistius' gloss on the preceding Aristotelian question at 215a20. *hôs te* suggests either that the text originally mirrored 215a20-2 more closely (as Spengel thought; see Schenkl's apparatus criticus at 130,4), or that it has intruded from someone's attempt to reproduce the Aristotelian text.

294. This sentence is best taken as a question; i.e. replace the comma after *iskhuroteron* (130,6) with a question-mark.

295. At 130,6 a colon rather than a stop should follow *logos*, given the proleptic phrase *epi touto*.

296. 130,13-17 = *SVF* 2.553. As Simpl. *in Phys.* 671,4-12 (= *SVF* 2.552) shows, this criticism originated with Alexander. See also Alex. *ap.* Simpl. *in Cael.* 286,6-10. For a Stoic response to the Peripatetic rejection of the claim that centripetal motion ensures stability in an infinite void see Cleomedes, *Caelestia* 1.1.91-5; see further Algra (1) and Furley (2). As for the 'cohesive *hexis*' Cleomedes, *Cael.* 1.1.98-9 argued that it ensured the non-dispersal of the cosmos. This *hexis* was the macrocosmic equivalent to the force that unified individual substances, and resulted from the motion of *pneuma* in the Stoic continuum. Hence *sunekhousa* cannot be translated 'containing' in the phrase *sunekhousa hexis*; that suggests an affinity with the static Aristotelian notion of place rather than the dynamic ('cohesive') continuum of Stoic physics.

297. They are so in the sense that they are either elements, or, as we see in what follows, qualitatively simplified compounds.

298. On the role of differing shapes in connection with the speed of bodies through mediums see the excursus below at 133,2-10.

299. In this paragraph (cf. n. 304), Themistius simplifies Aristotle's more formalised demonstration at 215a31-215b10.

300. This may be a reference to the *de Caelo*.

301. At 131,1 for *ison* I read *auto*. The received text has 'the equal weight' (*to ison baros*), but the point is that in this example the motion of a *single* heavy body is being analysed, as opposed to two different bodies of unequal weight discussed in the preceding paragraph. Thus instead of 'the equal weight' (equal to what?) we should expect to read 'the same weight' (*to auto baros*), as we do in a similar context at 131,20 below.

302. Literally 'heaviness' (*baros*), but in English 'the same heaviness' makes no sense, and so 'weight' has to be used, but without, of course, the accompanying notion of gravity.

303. If the volumes (or 'extensions') of the different mediums were disproportionate, then this body could go through a volume of water more quickly than one of air; this would not happen where identical volumes of each are involved.

304. Themistius creates a reader-friendly version of this text, in that he eschews Aristotle's proportion sums with their variables (215b22-216a4). These must be the 'additions' he refers to at 132,5 below as justification for an excursus that anthologises material from earlier commentaries.

305. At 131,14 the stop after *ho khronos* should be a comma, and the comma after *kineitai* should be moved to follow *diastêma*. I translate *diastêma* as 'spatial extension' to maintain continuity with the usual translation for this term as 'extension'.

306. Comment is omitted on the summation at 216a21-6.

307. At 131,28 Themistius uses *megethos* in its mathematical sense of an extension between two points, i.e. a length.

308. The syntax of this section is perhaps improved if the two complementary clauses at 131,29-30 and 30-1 in which specific numerical amounts for time and distance are assumed are bracketed as parentheses in their sentences.

309. See Themist. *in An. Post.* 1,16-17 on Aristotelian obscurity as the justification for his paraphrastic method.

310. 'Commentaries' here translates *exêgêseis*, i.e. explications.

311. The use of *epigelân* here (132,25) was probably inspired by its use in two memorable contexts in Plato's *Phaedo* (62A8; 77E3) to describe Cebes' reaction to Socratic claims.

312. Schenkl's orthography needs correction at 132,28 where *kisêreôs* should be read; i.e. with a single, not a double, sigma.

313. At 133,1 I have placed a stop after *kinêthêsetai* so that *hôste* can introduce a conclusion for the whole preceding paragraph, not just for the preceding sentence. Cf. n. 219, on 124,1 above.

314. At 133,1 I read *têi hupothesei* for *tên hupothesin*. The latter could be a phrase of respect, but a dative complementing *adunata* seems more plausible.

315. The contrast here is between a wafer-thin piece of iron and objects that we might call a pellet and a bar.

316. The verb *epokheisthai* (133,6) is used for vehicular transportation, e.g. for horses.

317. Here I emend the *lectio facilior allôs* ('in other ways') at 133,8 to *oxeôs* ('in a sharp way'). Without this change the explanatory clause that follows makes no sense; it cannot refer to just any non-blunt shape. Given his tendency to echo Platonic texts, Themistius may be reflecting *Timaeus* 61E3-4 where there is reference to fire's angles with which it cuts 'sharply'.

318. This is an epitomised version of an argument that Alexander had used. It is reported by Simplicius, *in Phys.* 679,12-37 (= Usener, *Epicurea* no. 279).

319. Themistius offers no equivalent to the summary at 216a21-6.

320. Cleomedes, *Caelestia* 1.1.33-8 also identifies this case, although less precisely, in the course of an argument for the existence of void as an extension distinguishable from, but always occupied by, body. At 1.1.36-8 he seems to think that the exiting air will be perceptible, without the use of special instruments, although he notes that a sound is more likely to be made when the neck of a container is narrow.

321. At 133,24 the *glôssai* mentioned are the reeds within the musical instruments. Since the word can also mean 'tongues', a confused text has produced the phrase 'the *glôssai* of flute-players (*aulêtôn*) and trumpets'.

'Flute-players' must be emended to 'flutes' (*aulôn*), since the sound is made by having the air that exits the clepsydra, in effect, play the adjoining instruments, which are at that vessel's mouth, not in the mouth of a player. This seems preferable to changing *salpingôn* ('trumpets') to *salpinktôn* ('trumpet players') with Shorey, 448.

322. At 133,29 for *ekbasaio* (force out) read *embiasaio* (force in) with Shorey, 448. This example enlarges Aristotle's reference at 216a31 to a body being either compressed or displaced. For the latter Themistius has to envisage a second body being forced *into* the clepsydra (hence my emendation) and displacement occurring by the vessel breaking under the pressure.

323. See Sorabji (3), 76 for a protest against what he dubs Themistius' non-Aristotelian 'concoction' that different extensions cannot coincide, i.e. be in the same place.

324. Ch. 4 above, especially at 113,27-8 (cf. 115,8-9).

325. At 134,30 I delete *to* before *sôma*; the latter should be anarthrous, as at 134,27.

326. At 135,6 read *kakeino allou*, which, without the crasis, is Diels' text for Simpl. *in Phys.* 682,17. Schenkl printed *kakeinou allo*, out of fidelity to his favoured MS **M**, which he was pleased to find confirmed by one of the Simplician manuscripts. But **M** has merely tried to correct the unacceptable reading *kakeino allo* in the other manuscripts of Themistius. The reading adopted certainly better conveys the infinite regress proposed here (and implied at Arist. 216b15-16) than saying clumsily that 'the void will need another [extension], and [there will be] another [extension] for that one too'.

327. 216b17-21 is not found in any of the Greek commentators, and deleted in modern editions.

328. The adjectives *manos* and *puknos* are widely translated as 'rare' and 'dense', with related variants for the nouns and verbs (*manôsis/puknôsis*; *manousthai/puknousthai*) that describe the processes of acquiring these properties. But it is misleading to use 'dense' to describe one aspect of a spectrum the whole of which identifies degrees of density, and which also correlates that spectrum with expansion and contraction. So while 'rare' and its variants (like 'tenuous' for *leptos*) convey the expansion of matter into a less dense form of greater volume, in English at least 'compact' and its variants best convey the contraction or compression of matter into a denser form and lesser volume, and so will be used for both *puknos* and *pakhus*, and their variants. Cf. also 139,3 below where *pêgnusthai*, a verb standardly associated with contractive freezing, is substituted for *puknousthai*. The issue of why change in density is associated with change in volume cannot be pursued here, except to note that it is open to obvious counter-examples, such as the saturation of porous bodies, like the pile of ash in a kylix (see n. 266).

329. This final sentence (introduced by *kai gar*) seems to attribute to the proponents of the void the idea that all change is change of place, as we would expect. That is, *kai gar* cannot be introducing an example of change, but rather a reference to the form that *all* change has in a discontinuous physical realm. However, at ch. 7 above, when analysing the relation between void and change from an external Aristotelian perspective (rather than stating his opponents' argument, as he is here), Themistius assumed that change of place was only one type of change recognised by the proponents of void, and that others could be explained without it; see 126,16-22 above with n. 252.

330. *kineisthai* at 135,20 obviously refers to vigorous movement rather than some undefined motion.

331. Xouthos (DK33), who is mentioned only by Aristotle, is an obscure figure, who may have combined Atomist and Pythagorean doctrine.

332. The brackets here introduce material that in a modern text would probably go into a footnote as a digression; they are not intended as a recommended punctuation for the Greek text.

333. This clause is Themistius' comment, derived from a later text in this chapter (217a15-16 and 18-20) where Xouthos is not mentioned. Having reported that Xouthos rejected the mutual replacement of bodies, he now notes that it can occur not only in a circle (i.e. not only among the circular-moving heavenly bodies?), but in a straight line, as Aristotle says in the later text. See also Simpl. *in Phys.* 683,28-30.

334. I have repunctuated 135,22-4 so that *hoson – hepetai* is bracketed as part of an ongoing sentence, with a second coordinating clause beginning with *ginetai d'oun* (line 23). For the correspondence of *men* (here at 135,22) and *d'oun* see Denniston, 460-1, and also cf. 160,5 below (n. 528).

335. Simpl. *in Phys.* 683,36-684,3 criticises Themistius' reading of this passage in the example that follows. He thinks that Xouthos' point (as reported by Aristotle) was that, if there is no intracosmic void, then when a spoonful of water turns into air, there will be a greater volume of air than there was of water, and that this will have to be compensated for by air somewhere else in the cosmos condensing into water. Themistius' idea that a spoonful of water will yield a spoonful of air is, as he says, impossible, but, more importantly, the wrong paradox to attribute to Xouthos. Simplicius' reading of Arist. 216b26-8 is surely correct, since at line 27 what is said is that '*air* comes to exist from a spoonful (*kuathos*, a twelfth of a pint, or six per *kotulê*; cf. n. 471) of water' not 'a spoonful of air'. Moreover, Aristotle himself, at 217a13-15 and 16-18 (paraphrased by Themistius at 137,18-21!), handles this issue as Simplicius urges.

336. At 136,1 I have deleted *kenon* and accepted MS **W**'s emendation *heloito* for *eroito* (cett.). That is, I take the reference of *toulatton* to be *to pan* (135,28), and this clause to be an account of why the All must bulge. *kenon* is a premature specification of what the All bulges into; that is only introduced by the disjunctive argument at 136,2-3. Also the phrase *toulatton kenon* is inappropriate, when the argument needs a phrase referring to the absence of void. As for *eroito*, it introduces too tentative a claim; i.e. if you concede that the All swells, then you will not 'ask about', or 'inquire after', an addition to it, but appropriate one. The one sense of *hairesthai* that might conceivably apply (see *LSJ*, *eromai*, 2) is 'learn of', but that seems unlikely here.

337. At 136,1 read *kumaneî* (future tense) for *kumainei* (present tense); note the use of the future tense in a parallel question at 136,6 below.

338. These are the arguments against place as extension (ch. 4 above), and against the separate void (ch. 8 above).

339. *kermatizein* (136,10), Themistius' term, probably borrowed from Arist. *Meteor.* 2.8, 367a11, where it is used of the disintegration of air into particles (*mikra*) prior to spontaneous combustion. Here, by analogy, the void is rendered particulate, and in that way 'dispersed'.

340. *sunkheisthai* refers elsewhere (114,3-4 above) to a unifying fusion, or conflation, of a body, so Themistius is praising his opponents for trying to minimise the presence of the void in a continuum, even if the project is doomed.

341. Themistius' version of Aristotle's terse analogy at 217a2-3 begs questions that are answered in Simplicius' version (*in Phys.* 684,35-685,2). The wine-skins (mentioned by Aristotle) must be inflated and float on water, as do the trawling nets (probably with floats) mentioned by the commentators; in this way they 'lighten' contents that would otherwise sink. The void would similarly endow bodies generally with such a capacity for flotation.

342. And we know that it does not have such a ratio from the argument at ch. 8 above, 131,11-26 (*ad* 216a13-21).

343. That is, the speed of void's upward movement could not be compared with that of air, or of any other body, moving in the same direction. See Bostock (2), 262 on 217a10.

344. Themistius omits comment on 217a15-16 and 18-20 (Waterfield, 300, *ad* 217a18-20, makes these texts consecutive), which concern the bulging that proponents of the void claim would be caused by its absence from the cosmos. But he has in fact already used this text at 135,22-4 above in connection with Xouthos' argument. Also, his paraphrase of 217a13-15 and 16-18 here is, as we have seen (n. 335 above), in conflict with his interpretation in that earlier context of 216b26-8.

345. The gloss on 'compacting' (*puknôsis*) here applies to both compacting and rarefaction (*manôsis*), since only jointly do they produce 'unequal volumes' (cf. 135,10-12 above). We could emend the text to add at 137,16 the words 'and rarefaction' (*kai manôsis*), which might have been omitted through homoioteleuton, or perhaps we can allow the same kind of looseness of expression granted above (see n. 103), when 'increase' included the complementary process of decrease.

346. The analysis of matter that follows can be compared with Alex. *Quaest.* 2.12, a briefer but still suggestive treatment; see Todd (1) and Sharples (3), 110-12.

347. At 137,27 I have ignored *epi* and translated over the lacuna inherited from the manuscript tradition. Whatever is lost is almost certainly a pair of coordinated clauses designed to re-emphasise matter's inseparability from form.

348. *metampiskhomenê* ('clothed') at 137,29, like *sunduastheisa* ('twinned') in the preceding line are both Platonic *hapax legomena*; see *Rep.* 569C3-4 and *Laws* 840D7 respectively. For several examples of Themistius' use of rare, or unique, Platonic vocabulary in his paraphrase of the *DA*, see Todd (6), Greek-English Index.

349. Themistius is more specific about the relation between matter and form than Aristotle. He also avoids (137,30) the standard term for form, *eidos*, and uses 'structure' (*morphê*) and the verb 'be shaped' (*skhêmatizesthai*).

350. The verb here, *apergazesthai*, has a teleological sense; cf. Themist. *in DA* 39,25.

351. Reading *hê autê hulê* ('the same matter') (with Arist. 217a28) for *autê hê hulê* ('the matter itself') at 138,11.

352. Schenkl's question mark at the end of this sentence (after *stoikheiois*, 138,17) is an error for a comma. Spengel used a colon.

353. At 138,17-18 we must assume *poioumen* as the main verb in this clause: literally, 'we cause there to be matter that is single'.

354. The Greek at 138,24-5 is awkward. There seems to be no precedent for construing *metalambanei* with an infinitive. However, the verb has a well-attested sense of 'receive in succession' (*LSJ* II.1) with a direct object, and so I have deleted *einai* at 138,25. I have also deleted *to* at 138,24; it seems a desperate attempt to create an articular infinitive with *einai*, but while that would give *metalambanei* a direct object specifying what is in potentiality, it would leave it without one for what is in actuality, unless *to* were supplied before *energeiâi* in line 24.

355. At 139,1 I read *all'allo ti*, with Simpl. *in Phys.* 689,10, for *all'hoti* of the Themistian manuscripts.

356. Literally 'in the [process] of something leaving off (*dialeipein*) and again not leaving off', i.e. 'by proceeding by fits and starts' (to use S. Leggat's felicitous gloss).

357. At 139,15 I read *tôi* instead of *to*, so that the articular infinitive phrase is taken as dependent on *en* in the previous line. This seems less strained than understanding the verb 'to be' and translating: 'but [more and less] are the [process of] total intensification and remission'.

358. The punctuation at 139,29-140,1 (i.e. brackets around *oute – manois* at 139,30) follows Ross (see on 217b20-1) in order to establish a dichotomy between separate (or actual) void (with its two manifestations as large and extracosmic, and as interstitial), and potential void. 'In effect' (*hoion*) is Themistius' qualification on potential void, not Aristotle's, but it is a reasonable addition, since this kind of void does not exist for an Aristotelian.

359. At 140,5 the claim that heaviness and lightness are causes *phoras ... kai kinêseôs* ('of motion and movement') is pleonastic. Arist. 217b26 contrasted *phora* with alteration, which he uniquely described as *heteroiôsis* rather than *alloiôsis*.

360. *dia tôn exôthen logôn*; i.e. arguments that are not technical. On the extent to which such arguments (which Ross identifies as 217b33-218a30; i.e. from here to 142,5 in Themistius' paraphrase) can be linked with Aristotle's 'exoteric writings' see Ross on 217b30-1. The presence of this phrase justifies Gigon's quoting 140,8-142,5 at T22.10 (167-8) of his collection of Aristotelian fragments.

361. At 140,13, instead of *kai mên kai eniautos* I read the text at Simpl. *in Phys.* 698,10-11, *kai hêmera mên eniautos* that Schenkl tentatively proposes. These examples of time-periods are used instead of the generic Aristotelian phrase at 218a1-2 *ho aei lambanomenos khronos* ('any given time').

362. Schenkl reasonably proposed moving this final clause (*oude – estin*) from the end of the paragraph (140,15) to this position. It anticipates 140,19-20 below.

363. The text here is incoherent, but the meaning is obviously the same as that of 218a2-3, with the modality changed from possibility to necessity. An implied main verb (*esti*, or *an eiê*) must be assumed to form a periphrasis with *ouk on* at 140,14, and *anankaiôs* (probably originally a gloss on *ex anankês*) in that line can be deleted.

364. A *pêkhus* is a cubit, i.e. the distance from an adult's elbow to the finger tips, or 24 'finger-breadths'.

365. These are discontinuous activities, *qua* activities. For example, battles are parts of a war, but a war does not consist of continuous battles.

366. At 140,15 above.

367. *katametrein*, the verb here (Aristotle just has the simple form *metrein*), refers to a measurement in terms of an invariable unit that can form a sub-multiple of the whole.

368. This legitimate addition to the argument at 218a6-7 is re-used at 141,18 below with similar reference to the character of the now.

369. *allo kai allo*, which is frequently used in the discussion of time. It literally means 'one different thing and another different thing', but such that the second thing is the first's immediate successor. I shall generally use 'one thing in succession to another' or close variants thereof.

370. Here (141,2) *enestanai* is used in a semi-technical sense to refer to present time (cf. 158,21 below). The month 'contains' the new moon since the day on which the new moon is first observed occurs during it. (Simplicius' example at *in Phys.* 698,10-11 is just of an undefined day 'contained' by a month.) Cleomedes, *Caelestia* 2.5.92-101 distinguishes senses of the word *mên* that include the time-period of a month, and the crescent moon.

371. See 157,10-29 below for this sense, there called 'strict' (*kuriôs*) and applying to the now that both divides time (the present from the future) and makes it continuous.

372. This additional example is an attempt to clarify the argument in spatial terms. The log is envisaged as simultaneously self-destructing and reconstituting itself.

373. Read *melloi* (MS **W**) for *mellei* at 141,15. Or else delete *an*, as Spengel did.

374. That is, not at the immediately succeeding now. Here Aristotle at 218a19-20 is a little clearer: 'Since [the now] did not cease to exist in a successive (*ephexês*) now, but in another one.'

375. I have replaced the question mark after *hen* in 141,28 with a comma, and moved it to follow *eniautos* (141,29), replacing the colon. Schenkl simply inherited Spengel's awkward punctuation, and failed to extend the question beginning at 141,27 to its conclusion.

376. This paraphrases Arist. 218a22-4: 'nothing divisible and limited has a single limit, whether it is continuous in one direction or more than one'. The omission of 'single' (*hen*) before 'limit' at 141,31 (noted by Schenkl *ad loc.*) so obscures the point that I have included it.

377. This was Pylos, and Neleus was the father of the elderly Nestor of Homer's *Iliad*.

378. At 142,7 I read *dêlon ti* (for *dêlon hoti*) with **W** (apparently) and Vettori. For the use here of *dêlon poiein* as equivalent to *dêloun* see Cleomedes, *Caelestia* 2.6.114.

379. i.e. the arguments presented to show that it does not exist.

380. Simpl. *in Phys.* 700,19-21 identifies these as the Pythagoreans (cf. Aetius 1.21.1 at *Dox. Gr.*, 318, and DK 58B33), and traces the doctrine to a misunderstanding of Archytas' claim that 'time is an interval in the nature of the whole'.

381. That is, 'being in' is defined differently when used of time and the heavens. See Arist. *Cat.* 1a1-6 on homonymy. In the discussion of the different sense of 'being in' at ch. 3 above, 108,6-20 (*ad* 210a14-24), 'being in time' was omitted (see n. 74).

382. Perhaps: All things are in time; All things are in the sphere of the heavens; therefore, the sphere of the heavens is time; i.e. All A is B; All A is C; therefore All C is B. But what follows is: All A is B and C, i.e. All things are both in time and in the sphere of the heavens, but in different senses, when 'in' is being used homonymously; see preceding note.

383. Simpl. *in Phys.* 700,16-19 records the consensus of earlier commentators that Plato (Ross cites *Timaeus* 39C5-D2) can be taken as identifying time with the revolution of the whole heavens.

384. Surprisingly at 142,20 Themistius has just *khronos* rather than *khronos tis* in this version of 218b1-2 where Aristotle says that 'even a part of the [celestial] revolution is a time'; clearly the reference of *khronos* is a specific time-period.

385. Here (142,22) and at 142,27 and 30, I take *to pan* as synonymous with *ho pas ouranos* ('the whole heavens') used elsewhere in this section.

386. Literally 'simple' (*haplous*).

387. I have deleted *autou* (sc. *tou khronou*; MSS **MW**) at 142,25. MSS **SL** have the reading *en autêi* (sc. *têi kinêsei*). But it cannot be claimed that the movement of the heavens is faster and slower 'than time', or, except redundantly, that it is so 'in movement'.

388. *Republic* 7, 514A2-6.

389. Themistius adds 'in the same place' (*en tôi autôi*) to his version of this text. For the ellipse *en tautôi* (sc. *topôi*) see Aristotle in the present book of the *Physics* at 209a6-7; 210b19; 211b11; 213b7 and 12; 214b6. What he means is that since there is a single time for the change that occurs in any of the multiple universes, then those universes form a single place at which multiple changes can occur at the same time, i.e. at multiple but identical times.

390. cf. Plato *Timaeus* 52D4, where Plato speaks of 'being, space and becoming' existing 'even before the heavens came into being'. The 'disorderly movement/motion' (as *ataktos kinêsis* is usually translated; I have used 'change' for consistency with the present context) is identified initially at *Timaeus* 30A5. For Aristotle's criticism of Plato based on the concept of the now see *Phys.* 8.1, 251b17-28 (paraphrased by Themist. *in Phys.* 211,26-212,9).

391. Themistius' remark is too brief for us to know whether he thought that in the *Timaeus* the orderly motion occurred in time. Simpl. *in Phys.* 704,13-29 quotes *Timaeus* 30A, and argues that time would exist along with the disorderly motion, if (as he himself did not accept) it existed prior to the orderly universe. See Urmson (1), 112 n. 62.

392. I have deleted *ou* before *perigraphetai* at 143,9; the point is that change *is* circumscribed by the content and location of what changes. The error may have a palaeographical explanation in dittography: *ou* iterates the final syllable (*-on*) of the preceding word, and given the similarity of upsilon and nu, could be written in an almost identical form.

393. The verb here, and in Aristotle, is *horizesthai*, traditionally translated 'be defined', but if unpacked etymologically means that something has its boundaries set by something else. Cf. 155,21-2 (*ad* 221a5) below where *horizesthai* ('be defined') is used to gloss *metreisthai* ('be measured').

394. The justification for translating the dative *khronôi* this way is given by Themistius himself when at 143,21 he glosses it by using the phrase 'by reference back to time' (*kata tên epi khronon anaphoran*).

395. By introducing a spatial extension (*diastêma*) here (it is absent from the guiding Aristotelian text at 218b15-17), Themistius ensures that in this sentence *kineisthai* refers to movement rather than to change in general.

396. At 143,27 I place a question-mark after *exeuroi. touto* is then proleptic before the asyndeton at *ennoêsômen*. Schenkl's punctuation would invite the translation: 'So let us conceive of how one might discover when and how we perceive time.' But what is involved here is forming a conception of how time itself is perceived (cf. 144,26 below for 'conception of time'), not of how it is to be discovered to be perceived.

397. A comma before *ei dokei* at 143,27 makes this an appropriately parenthetical remark to the reader or audience.

398. I have restored the *pote* at 143,28, where it was deleted by Schenkl, but have deleted the one at 144,1.

399. A slightly modified version of Homer, *Odyssey* 13,80, with the qualifying adverb 'most' omitted before 'like'.

400. This legend is reported by Aristotle. See Ross on 218b23-6.

401. At 144,13 a stop should follow *khronou*, since *houtô kai* is not introducing a correlative clause.

402. The activity to which Themistius is referring is undoubtedly thinking, which elsewhere he identifies as something that the human potential

Notes to pages 56-57

intellect causes us to tire from doing; see Themist. *in DA* 98,6 with Todd (6), 187 n. 20. Such an eventual result does not preclude the activity described here.

403. Aristophanes, *Clouds* line 1, the debt-ridden father Strepsiades' cry of anguish after a sleepless night of worry. At 144,19 by the word *hêmerai* (translated here 'days') Themistius means periods of twenty-four hours, and his point is that the burdensome nature of these full days is something of which awareness is intensified when night changes to day. He also takes the plural *tôn nuktôn* in the Aristophanic line to mean a plurality of nights, although in context it refers, by a standard idiom, to a single night.

404. The verb here (and at 144,26 and 145,1), *sunartasthai*, is used elsewhere (e.g., Sext. Emp. *Pyrrh. Hyp.* 2.111) to describe the relationship of mutual implication in conditionals, a sense appropriate here (cf. in particular 145,11-12 below), since to perceive time is to perceive change, and *vice-versa*.

405. On this passage and 149,4 ff. below see Moraux (2), 729 n. 170. Galen's views on time were probably criticised by Alexander of Aphrodisias, and so Themistius is probably extracting material from Alexander's lost commentary on the *Physics*. On Alexander and Galen on time see Sharples (1), 72-8.

406. Deleting the colon after *khronon*, and bracketing *houtô – Aristotelên* (144,25), ensures a symmetry between the two clauses beginning with *epeidê*.

407. Galen's point is that time's dependency on our changing does not prevent us from thinking of unchanging things, whereas, on Aristotle's view, things that are unchanging would have to change for us to think of them as being in time. See Urmson (1), 117 n. 171 on this and Simplicius' related discussion at *in Phys.* 708,27-709,12.

408. Below at 219a3-4.

409. *houtos* at 145,2 is used as a device for expressing contempt; see LSJ, *houtos* C.3.

410. It seems reasonable to take *en pollois* (145,2) as a reference to instances of Galen's Aristotelian exegesis, given Themistius' other criticisms of him in this book; see 114,7-20.

411. Ch. 10 above, 143,7-22 (*ad* 218b9-18).

412. Ch. 11 above, 143,25-144,23 (*ad* 218b21-219a1).

413. The terminology here is designed to unpack the phrase *ti kinêseôs* at 219a10, i.e. 'some element of change' (Ross, 386), or 'related to change' (Urmson).

414. The use of the preposition *epi* to govern 'extension' and 'magnitude' in the genitive case is designed to convey the extent of change, or the spatial equivalent 'over which' it extends. Hence I have translated it as 'over'. Urmson uses 'in' for related expressions at Simpl. *in Phys.* 710,23-4.

415. The language here is obscure, but the point seems to be that the extension persists (*hupomenein*) throughout the change, whereas the change itself is ongoing. This reinforces the idea of the extension as that 'over which' change occurs.

416. At 146,1 I read *epi* for *apo*, with MS Med. Laur. 85,14, cited by Schenkl.

417. The phrase *epi to peras* (146,1) means literally 'in the direction of the outer limit' and seems abrupt without a preceding verb, but I hesitate to supply one in the text.

418. The judges are presumably measuring the distance before a race, and the two points referred to here have to be marked on the oval stade-length course (*stadion*). Simplicius' explication at *in Phys.* 712,11-12 ('the before is

... the part nearest to the competitors, the later the end of the track', tr. Urmson) conveys the appropriate specifics: i.e. the start and finish, and the turning-point, here called 'the limit' (*to peras*). Cf. also Plotinus 3.7.9.62-3.

419. That is, they 'coincide' not in the sense of being superimposed, but in the sense of being two points that can be identified at the same time in a magnitude.

420. At 146,5-6 I have deleted the colon after *sunuparkhei*, and bracketed *hama – todi*, in order to clarify the syntactical coordination *epi men* (146,4)/*epi de* (146,6).

421. The received text at 146,6-7 would preposterously say that the before in change 'does not wait for the before in change'; obviously its destruction means that it does survive to await the *after* in change. I have therefore substituted *husteron* for the first *proteron* in 146,7. Simpl. *in Phys.* 712,15-16 has it right: 'in the case of change, when the before is destroyed, the after supervenes (*epiginetai*)' (my translation).

422. The text Schenkl printed at 146,8 is unsatisfactory. I have placed a stop at *ên* (146,7), and begun a new sentence with *ou mên*. Then in 146,8 I have deleted *hôsper ge*, which is probably a gratuitous repetition of *ge hôsper* in the preceding sentence, and replaced the colon after *tropon* in 146,10 with a comma. This admittedly tentative resolution at least has the merit of conveying the obviously intended sense that continuity is differently defined for a static extension and for a dynamic process of change.

423. Themistius' strategy with this text is to paraphrase it and then, for pedagogical purposes, introduce a specific example (147,8-23) to illustrate its assertion that recognising the passage of time involves perceiving the before and after in change.

424. At 147,8 Schenkl could not understand the sentence *alla tina thesin*. There should probably be a lacuna posited here to complete what was originally a question (*tina* being an interrogative adjective) that will then be addressed by the analysis of writing on a roll that follows. That question must have been an invitation to show how divisions in change resemble divisions in a line by points since this is immediately taken up at 147,10-13. My insertion here is not intended as an emendation but just an *ad hoc* device to ensure a smooth translation.

425. i.e. changes position; to use 'moves' for *kineisthai* at this point would confuse the argument.

426. At 147,14 I read *tis* for *dis* (cf. the use of *tis* just below at 147,16). It supplies an otherwise missing subject for the main verb in this clause. *dis* would anyway be redundant with a verb that already implies division 'into two'.

427. This dialogue suggests a classroom situation, with a papyrus roll serving as the 'blackboard'.

428. The *selis* is a column on a papyrus roll. For another use of writing to illustrate a philosophical theory (the theory of the potential intellect) see Themist. *in DA* 100,20-2.

429. At 147,21 for *proteron* read *to men proteron* (corrector in MS **L**).

430. For *houtô legein* (148,5) I read *houtô legein ekhei* (MSS **WSL**). Schenkl's defended *houtô legein* (the reading in his favoured MS **M**) as parenthetical, i.e. in the sense of 'so to speak'; that seems implausible in general and in this context.

431. Themistius' paraphrase of 219b3-5 omits the premise supplied here.

432. That is, to the definition of time at 148,17.

433. This example suggests a classroom situation in which a blushing

teenage (i.e. still growing) student has perhaps arrived late (i.e. is moving to his place in class).

434. A *khoinix* is a dry measure, a *khoeus* a liquid one. In terms of the *kotulê* ('half a pint'), the *khoinix* is four (the traditional diurnal corn ration), the *khoeus* twelve.

435. What is contrasted with 'essence' (*ousia*) here is *ergon*, which seems to mean the numerical product of the activity of counting units in a discontinuous series.

436. Themistius is echoing the description of Hector's spear at *Iliad* 6.319 (cf. 8.494). Since a *pêkhus* is twenty-four adult finger-breadths (the distance from elbow to finger tips), an eleven-*pêkheis* spear shaft (*doru*) would be around sixteen feet.

437. See n. 405 above, and Sharples (1), 78 on Alexander's reaction to this argument.

438. At 219a14-16.

439. In other words, the definiendum cannot appear in the definiens. So while Aristotle offers us 'Time = df. the number of change in respect of before and after', Galen proposes 'Time = df. the number of change in respect of time'.

440. At 149,23 I replace the question mark after *khronoi* with a comma, and the stop after *pantakhou* with a question mark; this produces smoother syntax.

441. Arguably a stop rather a comma could be placed after *lambanêtai* (149,26), but in either case the isolation of this problem as a separate text is justifiable.

442. The problem and solution here is Themistius' expository device for dealing with Aristotle's claim (219b12-15) that the now is the same and different in different senses, and he has made the illustration of Coriscus being different in different places (219b19-21) into part of the solution offered.

443. There may be humour intended in the picture of Socrates (who has replaced the Coriscus of the text) being in the Lyceum, the site of Aristotle's school.

444. This position (alluded to by Aristotle at 219b20-1), for which no Sophistic source has been traced, would involve treating all predications alike, and as essential definitions, so that once incompatible predicates were applied, the subject of them could not be the same. So here the claim that Socrates is 'another' (*heteros*) means that he is two different persons in the Lyceum and the Agora.

445. At 150,3 *ê* introduces the answer to the question raised in the problem at 149,27. So the question mark at 150,6 after *nun* can be eliminated, and the sentence allowed to run on to *logôi* in 150,7.

446. *Iliad* 8,186. The 'now' that Hector uses has inferential force in identifying a situation in light of the past, and so is an example of the now in its 'extended' sense, which can include the temporal environment of the now in its strict sense, as the divide between past and future; see below 157,30-32 (*ad* 222a20- 2).

447. See above 145,19-28 (*ad* 219a10-14).

448. In the following section (150,12-151,23) the verb *pheresthai* (and noun *phora*) are used to characterise that which time follows; they are translated as 'move' and 'motion', and seem deliberately employed to narrow the focus of the argument. Where *kinêsis* is used it is synonymous with *phora*, and can be translated 'movement'.

449. At 150,25 I read *to gar arithmêton auto* with Simpl. *in Phys.* 723,33, in preference to *auto gar to arithmêton auto* (Schenkl from the Themistian manuscripts). The iterated *auto* is unacceptable.

450. *en huparxei* (150,28) clearly carries a stronger force than 'in existence' since it refers to being a substance; we could have expected *en hupostasei*.

451. At 150,27-9 it is best to put both the explanatory clauses in brackets, not just the first. So at line 28 I have deleted the stop after *nun* and bracketed *en huparxei – tode*. I have also added *ti* after *tode* in line 29 to parallel this expression in line 27.

452. In justification of this translation of *peras*, note that it is a gloss on *teleutê* at Arist. 220a11 (and 13). Themistius himself uses *teleutê* at 151,19 below.

453. Here the verb is *kineisthai*, which has to be taken as synonymous with *pheresthai*, and translated as 'move' rather than 'change'.

454. Clearly *dis* ('twice') at 151,19 requires an accompanying form of the verb *lambanein* ('identify'), as it has at its two other occurrences in this context, at 151,18 and 23. I therefore read <*lambanomenon*> *dis*, and surmise that the exclusion of this qualifying participle was due to homoioteleuton, given the terminal syllable of the immediately preceding word *sêmeion*.

455. See above 140,15-23 (*ad* 218a3-14).

456. At 151,27 I read *tôi nun* (dative case), with MS **W**, for *to nun*. Otherwise *diaireisthai* would have to be construed abnormally (though not impossibly; see Arist. *PA* 642b5) as the middle voice.

457. The Aristotelian text isolated here is problematical; see Ross *ad loc*. Themistius is perhaps offering more of a reconstruction than a paraphrase, and Ross may not be justified in attributing a different text to him because he uses the status of the now as a limit of change in contrast with its status in relation to time.

458. The sentence at 152,1-4 introduced by the particle *ê* is an answer to the question at 151,26-7, not another question, as Schenkl makes it. Why else would the next sentence draw a conclusion (*dia touto ara*, 152,4)? So replace the question mark after *peras* (152,4) with a stop.

459. The way that this triad of changes is gratuitously introduced without conjunctions makes it suspect as an obvious gloss on *tôn pollôn kinêseôn* (152,11).

460. I have omitted *hoion* at 152,16; there seems no satisfactory way of construing this gratuitous qualification on *hê duas*. The Aristotelian text has *haplôs* ('without qualification'). On the Greek belief that two was the first number, since the unit was not itself a number but a limit on number, see Ross on *Phys.* 220a27-32, and Wagner, 577.

461. Here Themistius creates a *quaestio* to give the abrupt opening of this chapter in the Aristotelian text some dialectical context; Aristotle begins by just distinguishing different senses of 'number'. At 152,17 *ê* then has to introduce the answer to his question, not pose another, as Schenkl followed Spengel in assuming; their question mark after *grammôn* at 152,19 should therefore be replaced with a stop.

462. At 152,21 I have deleted the colon after *to elakhiston*, and bracketed *elakhiston – toiauta* (152,21-2), replacing the stop after *grammai* with a colon, and followed the clause with a comma. This coordinates *kata men to plêthos* and *kata megethos de*. By also deleting the colon after *hexei* and bracketing *hekaston – diaireton* (line 23) balanced explanatory clauses are created for each coordinated clause in this sentence.

463. At 152,24 I delete *hai duo hôrai* as an obvious intrusive gloss presumably from the reference to 'two hours' in line 27. Also in line 24 I place a comma rather than a colon after *estin*, delete *de* after *hexei* (with MSS **WBL**), bracket *ou – touto*, and delete *de* in line 26. This respects the coordination *katho men/katho de* (152,24-5).

464. cf. ch. 10 above, 143,18-21 (*ad* 218b17-18).

465. This translates Torstrick's proposed supplement *hôi men oun arithmoumen* for the lacuna he posited, and Schenkl retained, at 153,5.

466. Here Themistius has over-simplified 220b9-10: 'this [number as what is counted] turns out always to be different before and after, because the nows are different'.

467. At 153,9 I read *palin kai palin* with MS **W** of Themistius (cf. Simpl. *in Phys.* 733,1) and Arist. 220b13-14. The reference to repetition is essential.

468. Schenkl's suggestion (app. crit. 153,8) that hereabouts Themistius may be following Eudemus (at Simpl. *in Phys* 732,26-733,1; Wehrli fr. 88) seems unjustified.

469. Themistius' conclusion, when set against the two preceding sentences (corresponding to 220b9-10 and 13-14), supports Bostock's complaint (Bostock [1], 157) about Aristotle's lack of distinctions in the senses of 'time', since we have an apparent contradiction between non-recurrent time (the different nows [cf. n. 466 above] that are before and after), and the recurrent time manifested by the seasons.

470. See Simpl. *in Phys.* 733,16-18 for the same terminological accretion to the Aristotelian text to create this contrast: 'Time measures change directly (*proêgoumenôs*) through being its number and by defining it in respect of the before and after, while it is also measured in turn by it somehow (*pôs*) incidentally (*kata sumbebêkos*)' (my translation). The qualification 'somehow' could be justified because this is an extended sense of 'incidentally'. Thus a lot of change is necessarily measured by a lot of time (= 'direct' measurement); but a lot of time is not necessarily measured by a lot of change, but will always be measured by *some* change, so that in this special sense the 'counter-measurement' (*antimetreisthai*), or reciprocal measurement, of time by change is 'incidental'. Unfortunately this implication is not pursued, and the conclusion at 154,3-4, that time is a large quantity if change is, is left unqualified.

471. A *medimnos* is a measure of corn, a *kotulê* a liquid measure of about one-half of a pint. The metal involved here is presumably molten.

472. At 153,22 I delete *kai ton hena anthrôpon*. The phrase would be tolerable only if it were, like the references to horses, iterated (i.e. 'one person and another person'), as perhaps it should be (*homoioteleuton* would explain the omission) to create a second example here.

473. Ch. 11 above, 145,11-15 (*ad* 219a8-10).

474. A *pêkhus* was a small measure ('twenty-four finger-breadths'), of which there were 400 in an ancient stade.

475. This 'change' (*kinêsis*) is a change in the circular motion of the Sun; this is more precisely conveyed at 163,20 below.

476. *kai ho mên kai to etos* (154,7) is implausible as a sentence; it is probably a gloss that has intruded here.

477. At 154,9-10 the parenthesis should include the clause *ou – kineseôs*, which cannot be linked to the preceding clause. At 154,9 *allo* is comparative in force, and is complemented by *ê*.

478. At 154,15 I supply *tôn* before the meaningless *autôn*.

479. In sense (ii-a) above.

480. In sense (ii-b) above.

481. 154,21-3 is best punctuated with the two explanatory clauses that

supplement each of the correlative clauses made parenthetical; so bracket *hôrismenos – horismos* (line 22) and *kai gar – tis* (line 23), delete the cola after *arithmos* and *horismos* (line 22), follow *horismos* with a comma, and delete the colon after the first *khronos* in line 23.

482. That is, in sense (i) above.

483. This is a reference to Aristotle, *Physics* 8.8.

484. This translates *hexeis*, used here in this specific sense.

485. At 156,1 the syntax requires *epei de* (achieved by dividing *epeide*, Schenkl's report of MS **M**) instead of *epeidê*.

486. This verb, carried over from the previous sentence, is actually supplied in MSS **WBL**. Without it the syntax is unusually elliptical, but the meaning is still clear.

487. The difficult text at 156,22-3 is best approached through a close parallel at Simpl. *in Phys.* 744,6-8: 'For being for *this* [sc. rest] does not consist in the transition (*metabasis*) from something before to something secondary (*deuteron*) and subsequent (*met'auto*), as it does for change' (my translation). Clearly this is the thought that Themistius is trying to convey, and it can be extracted from his text as it stands, if the awkward position of *to einai* vis à vis the complementary phrase *kai tautêi* can be accommodated. I have offered, without complete confidence, a translation that parallels Simplicius' version, and have also bracketed this clause (*ou – husteron*), while deleting the colon after *ou* in line 22.

488. This phrase must be understood as a generic untensed phrase referring to an event that can occur at any time. Themistius' choice of this example (there is none in the Aristotelian text) is undoubtedly dictated by the fact that the Persians in his lifetime were as much a threat to the eastern Empire as they had been in classical times to Greece. His *Oration 1* (delivered *c.* AD 350), in fact refers to the Persian wars of the Emperor Constantius.

489. In line with the standard punctuation at Arist. 222a6 I have created a new sentence after *khronôi* at 157,7 beginning *ou toinun*. It would be unusual for *toinun* to be a coordinating particle for an additional clause.

490. On this distinction, often applied to place, see n. 45 above. Cf. also Arist. *Phys.* 6.3, 233b33-5 with Themist. *in Phys.* 189,21ff.

491. Ch. 11 above, 150,30-151,23 (*ad* 220a4-18).

492. 'Held together' (*sunekhetai*), or 'made continuous'; cf. 151,4-5 above for the periphrasis 'makes continuous' (*sunekhes poiein*). The etymologically related adjective *sunekhês* (literally 'holding-together'; e.g., 158,2 below) is translated 'continuous'.

493. cf. 147,17-18 above.

494. I have placed a colon after *diaphora* at 157,27 to replace the comma that Schenkl mechanically inherited from Spengel 333,18.

495. This sentence is Themistius' version of 222a19-20. There Aristotle says: 'The division and the unification are the same thing, and pertain to the same thing, but the being is not the same'. This remark is meant to cover both the point and the now; they are the reference of 'the same thing' (see Ross on 222a19-20). Themistius, however, drives a wedge between the point and the now by suggesting that the being for dividing and unifying differs for each of them, presumably because the now unifies and divides potentially.

496. At *in DA* 110,34-6 Themistius notes that 'the extended now' (*to en platei nun*) is a necessary condition for thinking what he there calls 'the undivided now' (*to ameres nun*), which he refers to here as its strict sense.

497. See LSJ, *nun* I. 3 and 4 for these two senses.

498. To fit with the preceding sentence this Flood has to be a future event.

Notes to pages 67-68   107

The Aristotelian text at 222a23, however, has the same verb, *gegone* ('has come to be') in the clause referring to the Flood as in that referring to the past event, the Trojan War. Ross *ad loc.* follows Torstrick in deleting it. The future tense at 158,5 below suggests that Themistius might have welcomed an implied future tense here too, and the translation reflects this.

499. The standard translation for this indefinite temporal modifier is 'at some time' (Hardie and Gaye; Hussey; Waterfield), but, as Pamela Huby has pointed out to me, this undermines its definition in terms of time. I have therefore experimented with a neologism; i.e. if accented *pote* means 'when?', unaccented *pote* needs only a minimal marker of indefiniteness. Ross' 'some-time' (in his summary of 222a24-9 at 390; copied in Urmson [1]) did not therefore go far enough.

500. The Aristotelian text (at 222a25) refers only to 'the now that is before', which could be taken to mean the strict sense of the now discussed above: 'the first kind of now' (Waterfield), '[now] in the former sense' (Hussey). But Themistius uses the phrase not as such a reference back, but in the chronological sense defined in the preceding paragraph, and contrasts it, along with 'the now that is after', with the present now. In relation to these parameters something happening 'somewhen' is 'bounded' by two nows that are defined in relation to the present now.

501. In both clauses of this sentence Themistius qualifies Aristotle's 'time' with 'identified' (*lambanomenos*), thereby spelling out that (in Ross' words, *ad* 222a29) this is not 'time as a whole, but every particular period of time'.

502. I have punctuated 158,10 so that it reflects the standard punctuation of Arist. 222a29-30. Thus the question as to whether time will give out, is followed by a negative response in the form of a leading question introduced by *ê* (so construed by Hardie/Gaye and Waterfield), and so a question mark should be placed after *pote*. Aristotle's answer is 'Surely not so, if change is everlasting'; Themistius' is 'Not so, since [there is then] not even change' (*ê ou, epeiper oude hê kinêsis*). If this seems too elliptical, *aei esti* would have to replace *oude hê*, as Schenkl saw.

503. The sentence at 158,10 introduced by *ê* must be a direct response to the preceding questions (as at Arist. 222a31), not the additional question Schenkl makes it. The question mark after *khronos* at 158,11 must be replaced by a stop.

504. Themistius has just reproduced the Aristotelian text, and not spelt out its implications. For the missing paraphrase we can use Simpl. *in Phys.* 751,4-5: 'as in change, the same [time] will recur over and over, but numerically ever different while the same in kind (*eidei*)' (tr. Urmson); or Ross (*ad* 222a31-3): 'since what recurs is not a numerically but only a specifically identical [change], the same is true of time'. Themistius has covered the general issue of the recurrence of the now at ch. 11, 151,1-23 (*ad* 220a4-18).

505. Themistius omits Aristotle's example (222b2-4) of a circle's simultaneously concave and convex shapes being analogous to the now's being simultaneously a beginning and an end.

506. That is because, as Aristotle explains (222b5), the now would be both the beginning and the end of the same time.

507. At 158,14 read *hama gar an eiê* to reflect Arist. 222b5-6. Schenkl's omission of *an* with an optative, to judge from his apparatus criticus where he reports the correct and majority reading first, has to be an error.

508. Translators have problems with *êdê* because it points to both the past and the future, and there is no happy single English equivalent, short of the word play that I have adopted here. Waterfield tries 'soon' but cannot apply it to the past and falls back on 'already'. Hussey has 'just' which can certainly modify past and future tenses, but produces implausible English for the example applied to the future by Aristotle: ' "When are you taking a walk?" "I'm just taking it." ' In that case the walk would no longer be a future event, since it would 'already' be under way, rather than being on the point of being undertaken, as 222b9 (= Themist. 158,17) requires by saying that 'the time in which he is going to (*mellei*) walk is near'.

509. It is difficult to know whether to translate *mellei* as 'it [the walk] is going to happen', or 'he [the walker] is going to walk'. Perhaps Themistius' text should be adjusted to follow Aristotle's (222b8) so that the question here becomes 'When are you going to walk?' (*pote badizeis;*).

510. Themistius changes the present tense in Aristotle (*badizeis*, 222b10) to a perfect tense, in order to anticipate the response in Aristotle (*êdê bebadika*, 222b10-11, 'I have already walked'). In that way, he makes the questioner aware that the walk is complete, but uncertain as to when, whereas Aristotle has the questioner imply that the walk is not yet taken; i.e. the present tense is prospective ('When are you going for a walk?', Waterfield).

511. The verb is *existasthai*, regularly used of an abrupt displacement from a prior condition. Waterfield's 'a shift' is better than Hussey's 'removal from a previous state', which fails to capture the required abruptness.

512. This is Simonides (c. 556-468 BC; the lyric and elegiac poet from Ceos), fr. 19 Bergck (3:395, no. 1123).

513. 'By the agency of time' (*hupo khronou*), derived from its single use by Aristotle in this context (cf., however, ch. 11, 221a30-222b1 = Themist. 155,10-17) at 222b25, is justifiably translated in personifying terms, since Aristotle also speaks (222b20) of time being 'responsible' (*aitios*). Urmson (1) uses 'through time' (*ad* Simpl. *in Phys.* 753,10 and 754,30-1), which misleadingly implies duration of time.

514. Aristotle's is the only reference made to this individual, whom he calls a Pythagorean (see DK 26). On the basis of a report of Simonides' statement by Eudemus (fr. 90), in which there is a reference to an occasion when a person 'being present' (*parôn*) contradicted Simonides, Simplicius (*in Phys.* 754,7-17; 9-13 cited in DK) thought that there was no sound basis for identifying a named individual at all.

515. Ch. 12 above, 155,11-17 (*ad* 220b30-221b3).

516. i.e. time. Themistius has expanded 222b21-2 to make change as well as time the incidental cause of coming into being.

517. <not> at 159,8 is supplied as an emendation, suggested by Pamela Huby: i.e. read <*ouk*> before *arkei*. *A fortiori* Simpl. *in Phys.* 754,26 should be similarly emended, as it tacitly is by Urmson (1), 168 through the translation: 'time alone is not sufficient etc.'. But the combination *ou monon* that he separates here means 'not only' as a single modifying expression counterbalanced by *alla*, and so the whole clause can be negated only by inserting a further negative.

518. Themistius is almost certainly using 'putrefy' (*sêpesthai*) with reference to Arist. *GA*, 3.11, 762a13-14 where it is said that 'nothing comes into being when putrefying but only when being concocted'. For Themistius' engagement with the issue of spontaneous generation discussed in that text see Brague, 35-7, and in particular Henry; I am grateful to Devin Henry for alerting me to this area of Themistius' interests by showing me an advance copy of his paper.

519. At 159,14 omit *tên* before *metabolên* (MSS **BL**).

Notes to pages 69-70 109

520. *sumbainei* at 159,13 (222b26) is intended to convey the idea of an incidental relationship between time and ceasing to exist. Urmson (1) 166 has 'happens contingently'.

521. Here *metra* refers to the limits within which evolution occurs, not just the measurement of such change.

522. At 159,16 (*monente Huby*) I read *en genesei* with MS **W** and Simpl. *in Phys.*755,2 ; Schenkl omits *en*.

523. In other words, when something ceases to be, it does so by virtue of natural causes; the time of the cessation is secondary or incidental.

524. Themistius has added the redundant verb 'change' (*kineisthai*); at 222b31 Aristotle simply leaves the verb 'to be' understood.

525. Themistius glosses Aristotle's *to hupokeimenon* (223a1), since the term normally refers to the substantial identity of something, as opposed to a description under which it may fall. Ross refers to *Phys.* 5.6, 229b29 for its similar use to mean the terminus of change; Themistius' paraphrase of the latter (*in Phys.* 178,6) omits it.

526. This reference to locomotion means that *kinêsis* and *kineisthai* must be translated 'movement' and 'move' in the lines that follow.

527. This further clause is Themistius' insertion. His point is that in terms of locomotion A can cover a distance faster than B, but B may be transformed in some other way before A. Perhaps if A and B are runners, B's complexion becomes flushed first.

528. At 160,5 I read *alla tauta* (**W**; *toutôn* cett.) *men exôthen* (with *exôthen* in the same sense of 'non technical' found at 140,9 above) followed by a comma, leaving *êkribologeisthô* to introduce a contrasting clause marked by *d'oun* (cf. n. 334 above). Otherwise the new sentence introduced by *hoti* at 160,5 lacks a main verb. Themistius' sentence *alla – en khronôi estin* (160,5-7) expands Arist. 223a4: *alla mên to ge proteron en khronôi estin*.

529. The comma after *husteron* at 160,7 should be deleted, and the gratuitous *estin* deleted after *en khronôi*. The clause then becomes an expansion of Arist. 223a4, *to proteron en khronôi estin*. Schenkl's punctuation of lines 5-7 seems a desperate attempt to make sense of a sentence without a main verb.

530. At 160,8 *to – mellontos* is a parenthesis, and should be bracketed, with the colon preceding it deleted, and the one following replaced by a comma.

531. Schenkl says that he derives this supplement at 160,9 from Simplicius, but it is at Arist. 223a7 (quoted at Simpl. *in Phys.* 756,30).

532. There is no comment on 223a8-13 where before and after are defined in relation to the past and the future and their distances from the now.

533. Boethus of Sidon was a Peripatetic philosopher of the middle of the first century BC; see Schneider. Moraux (1), 170-1 regards Boethus' comment here as a genuinely intended objection, not a dialectical manoeuvre. Huby shows that Boethus' discussion of time was, as Moraux suspected, rooted in his commentary on Aristotle's *Categories*; see Huby, 408 on the present text. Themistius was probably following Alexander; see Simpl. *in Phys.* 759,18-760,3.

534. Schenkl does not put the second clause ('just as ...') in quotation marks, but it is integral to Boethus' thought, and taken as such by Moraux (1), 171. Thus move the closing quotation mark from *arithmountos* (line 27) to *aisthanomenou* (line 28).

535. At 160,29 I supply *to* before *arithmêtikon* to balance the definite article before *arithmêton* in the next clause.

536. Or, it is implied, as an object of perception can without sense-perception occurring.

537. This seems to be a general reference to the role of the unmoved mover. In his paraphrase of Aristotle *Metaph.* 12.8, 1074a35-8, where the uniqueness of this mover is asserted, Themistius remarks that the diffusion of this mover in the celestial spheres involves 'desire'. For the text, based on the Arabic, see Brague, 104 at para. 15.

538. 'Product' translates *ergon* (161,8), which has to refer to the result of animals' activity, not the activity itself, as at 163,7 below, where it is a synonym for *energeia*. (I am indebted to Pamela Huby on this point.)

539. See Themist. *in DA* 100,24-6 on matter receiving divided qualities.

540. At 161,22 I have deleted *hepta* ('seven') and the comma that follows it. It interrupts the syntax, and almost certainly originates in a marginal or interlinear gloss.

541. At 161,22 I read <*epi*> *tôn kineseôn* to parallel *epi tou monadikou arithmou* (161,19-20).

542. Since what is counted are the changes, *arithmoumenai* should be read here (161,24).

543. At 161,24 read *kata tauta* (MS **ML**), with the latter accented so as to mean 'these' rather than in a crasis of *ta auta* (MS **WB**) to mean 'the same things'.

544. In the next three paragraphs Themistius prepares for the introduction of the circular motion of the heavens as the measure of time by extracting three problems from 223b12-21, but without mentioning circular motion, which Aristotle introduces at 223b12-13 and 18-20. In this way he is trying to impose some structure on a rather disorganised text.

545. Here Aristotle asserts that 'smooth' or 'uniform' (*homalês*) change is not found in the cases of alteration, increase or coming-into-being. Themistius recreates this remark as a problem for these changes, and then uses the whole text later in direct paraphrase at 163,14-16 below. Simpl. *in Phys.* 762,11-24 quotes Themist. 161,29-162,11, with some minor variants.

546. It has to, since the processes themselves have no component that represents change in terms of differing positions separated by an extension.

547. This text is itself a premise ('every kind of thing is measured in terms of some one thing of that kind – units in terms of a unit, horses in terms of a horse'), but Themistius uses it as a basis for a general problem about the unity of time when it is related to different entities. Simplicius thought this problem worth raising, and quoted Themist. 162,11-163,7 at *in Phys.* 765,32-766,19. Diels, the editor of Simplicius, relied on Spengel's text from which Simplicius' version deviates not only by providing variant readings, but by on occasion being more of a paraphrase. For a translation of the Simplician text see Urmson (1), 179-80.

548. 'Property' is used for *pathos* here, since the term is being used generically.

549. At 162,19 for *hôs* I read *hôste* with Simpl. *in Phys.* 766,6 to ensure an emphatic consequence. See n. 78 above on 109,15 for a similarly motivated emendation.

550. That is, we cannot conceive of the now as being simultaneous with itself.

551. Simpl. *in Phys.* 766,8-9 has a more expansive version of this conditional: 'If there are not many nows, but one and the same now, it [the now] will be separate and time will in no way belong to change.'

552. 163,1-7 is quoted by Simpl. *in Phys.* 766,13-19.

Notes to pages 72-73

553. *touto* (163,1) refers to the conclusion of the preceding problem. That is, if what is counted has no substantial being (the force of *hupostasis* at 163,1), then it is legitimate to revert to the earlier issue of time being a conception of our mind, and to criticise it.

554. At 223a21-8 above, though the aporetic context there makes any definite conclusion uncertain.

555. Schenkl marked the end of this sentence (163,5) with a stop rather than the required question-mark.

556. On Boethus see n. 533 above. On this particular text see Moraux (1), 171 and Huby, 408.

557. What follows is a paraphrase of 223b13-15. Themistius is saying that despite earlier suggestions that time is a conception of the mind, the immediate text attempts to ground it in another objectively defined time.

558. The basis for the problem at 161,29-162,11 above; see n. 544.

559. At 163,16 *hama* (tr. 'along with') is used in a logical rather than a temporal sense; the point is that celestial motions can be easily counted.

560. At 163,19 I have, like Spengel, omitted one of the iterated instances of *khronos*. Schenkl suggests in his apparatus that in light of Simplic. *in Phys.* 768,13. Themistius' text may originally have stated that the hour was a part of time, but it seems enough for him to say in contrasting clauses that the hour is *a* time, but specifically one that delimits a part of the celestial motion.

561. cf. 154,6-7 above. There may an implicit reference here to sun-dials (*hôrologeia*), for which the Greek term means literally 'hour indicators'.

562. Time is of longer duration than any given unit of time that measures a finite change; see ch. 12 above, 154,21-155,3 (*ad* 221a26-8). Themistius inserts this principle into the present context.

563. See ch. 10 above, 142,18-143,7 (*ad* 218a33-b1).

564. 224a2-17 receive no comment.

565. There were 400 *pêkheis* in a stade.

Bibliography and Abbreviations

ACA = *Ancient Commentators on Aristotle*, London and Ithaca NY, 1986-.
Accattino, P. and Donini, P. *Alessandro di Afrodisia: L'anima*, Rome and Bari 1996.
Algra, K. (1) 'The early Stoics on the immobility and coherence of the cosmos', *Phronesis* 33, 1988, 155-80.
───── (2) *Concepts of Space in Greek Thought*, Leiden 1995. *Philosophia Antiqua* 65.
Balleriaux, O. 'Thémistius et l'exégèse de la noétique aristotélicienne', *Revue de philosophie ancienne* 7, 1989, 199-233.
Bergck, T. (ed.), *Poetae Lyrici Graeci*, ed. 4, Leipzig 1882.
Blumenthal, H.J. (1) 'Photius on Themistius (Cod. 74): did Themistius write commentaries on Aristotle?', *Hermes* 107, 1979, 168-82.
───── (2) 'Themistius: the last Peripatetic commentator on Aristotle?', in G. Bowersock *et al.* (eds), *Arktouros: Hellenic Studies presented to Bernard M. Knox*, Berlin and New York 1979, 391-400. Reprinted with revisions in Sorabji (4), 113-23.
Bostock, D. (1) 'Aristotle's account of time', *Phronesis* 25, 1980, 148-69.
Bostock, D. (2) See Waterfield, R.
Brague, R. (trans.) *Thémistius: Paraphrase de la Métaphysique d'Aristotle (Livre Lambda)*, Paris 1999.
Browne, G.M. (1), 'Ad Themistium Arabum II,' *Illinois Classical Studies* 23, 1998, 121-6.
───── (2), 'Aristotle, *De Anima* 428b18-25', *Classical Quarterly* 49, 1999, 629-30.
CAG = *Commentaria in Aristotelem Graeca*, Berlin 1882-1909.
Cacciatore, P.V. 'La parafrasi di Temistio al secondo libro degli *Analitici Posteriori* di Aristotele', in Moreschini, 389-95.
Cherniss, H. *Aristotle's Criticism of Plato and the Academy*, vol. 1, Baltimore 1944.
Ciollaro, M.C. 'Osservazioni sulla *Parafrasi* di Temistio al *De Anima* Aristotelico', in Moreschini, 79-92.
Denniston, J. *The Greek Particles*, Oxford 1934; ed. 2, rev. K.J. Dover, Oxford 1950.
Dox. Gr. = H. Diels (ed.), *Doxographi Graeci*, Berlin 1879; repr. Berlin 1965.
DK = H. Diels (ed.), *Die Fragmente der Vorsokratiker*, rev. W. Kranz, ed. 6, Berlin 1952.
Furley, D.J. (1) (trans.) *Place, Void, and Eternity: Philoponus, Corollaries on Place and Void*, London 1991. *ACA*.
───── (2) 'Some points about Stoic dynamics', *Proceedings of the Boston Area Colloquium in Ancient Philosophy* 9, 1993, 57-75.

Gigon O. (ed.) *Aristotelis Opera: Volumen Tertium, librorum deperditorum fragmenta*, Berlin and New York 1987.

Gottschalk, H.B. (1) *Strato of Lampsacus: Some Texts*, Leeds 1965. *Proceedings of the Leeds Philosophical and Literary Society (Literary and Historical Section)* XI:6, 95-182.

―――― (2) 'Aristotelian philosophy in the Roman world from the time of Cicero to the end of the second century CE', *Aufstieg und Niedergang der Römischen Welt* II.36.2, 1987, 1079-1174.

Hardie, R.P. and Gaye, R.K. (trans.) *Physics*, in J. Barnes (ed.), *The Complete Works of Aristotle: The Revised Oxford Translation*, vol. 1, Princeton 1984, 315-446.

Henry, D. 'Themistius and spontaneous generation in Aristotle's *Metaphysics*', *Oxford Studies in Ancient Philosophy*, 2003 (forthcoming).

Huby, P.M. 'An excerpt from Boethus of Sidon's commentary on the *Categories*', *Classical Quarterly* n.s. 31, 1981, 398-409.

Hussey, E. (trans.) *Aristotle's Physics Books III and IV*, Oxford 1983.

King, H.R. 'Aristotle's theory of TOPOS', *Classical Quarterly* 44, 1950, 76-96.

Lautner, P. Notes at Urmson (4), 223-36.

Lettinck, P. (trans.) *Philoponus: On Aristotle Physics 5-8*, London 1994. ACA.

Lewis, E. 'Diogenes Laertius and the Stoic theory of mixture', *Bulletin of the Institute of Classical Studies* 35, 1988, 84-90.

LS = Long, A.A. and Sedley, D.N. *The Hellenistic Philosophers*, 2 vols, Cambridge 1987.

LSJ = H. Liddell, R. Scott, H.S. Jones, *A Greek-English Lexicon*, ed. 9, Oxford 1940.

Machamer, P.K. 'Aristotle on natural place and natural motion', *Isis* 69, 1978, 377-87.

Moraux, P. (1) *Der Aristotelismus bei den Griechen von Andronikos bis Alexander von Aphrodisias*, Bd. I: *Die Renaissance des Aristotelismus im 1. Jh. v. Chr.*, Berlin 1973. Peripatoi 5.

―――― (2) Bd. II: *Der Aristotelismus in I. und II. Jh. n. Chr.*, Berlin 1984. Peripatoi 6.

―――― (3) W. Kullmann, R.W. Sharples, J. Wiesner (eds), Bd. III: *Alexander von Aphrodisias*, Berlin 2001. Peripatoi 7/1.

Moreschini, C. (ed.), *Esegesi, Parafrasi e Compilazione in Età Tardoantica*, Naples 1995.

Morison, B. *On Location: Aristotle's Concept of Place*, Oxford 2002.

Mueller, I. 'Aristotle's doctrine of abstraction in the commentators', in Sorabji (4), 463-80.

Nauck, A. (ed.), *Tragicorum Graecorum Fragmenta*, ed. 2, Leipzig 1889.

Nutton, V. 'Galen's philosophical testament: "On my own opinions",' in J. Wiesner (ed.), *Aristoteles: Werk und Wirkung, Zweiter Band: Kommentierung, Überlieferung, Nachleben*, Berlin and New York 1987, 27-51.

Obbink, D. ' "What all men believe – must be true": common conceptions and *consensio omnium* in Aristotle and Hellenistic philosophy', *Oxford Studies in Ancient Philosophy* 10, 1992, 194-231.

Penella, R.J. (trans.), *The Private Orations of Themistius*, Berkeley 2000.

Philop. *in Phys.* = H. Vitelli (ed.), *CAG* 16-17, Berlin 1887, 1888.

Pignani, A. 'La parafrasi come forma d'uso strumentale', *XVI Internationaler Byzantinistenkongress*, II.3, Vienna 1982, 21-32.

Radt, S. (ed.), *Tragicorum Graecorum Fragmenta*: vol. 3, *Aeschylus*, Göttingen 1985.

Rashed, M. 'Alexandre d'Aphrodise et la "Magna Quaestio": rôle et indépen-

dance des scholies dans la tradition byzantine du corpus aristotélicien', *Les Études Classiques* 63, 1995, 296-351.
Renehan, R. *Studies in Greek Texts*, Göttingen 1976.
Ross, W.D. (ed.) *Aristotle's Physics: a revised text with introduction and commentary*, Oxford 1936.
Sambursky, S. *The Physical World of Late Antiquity*, London 1962.
Schenkl, H. (ed.), *Themistii in Aristotelis Physica paraphrasis*, Berlin 1900. *CAG* 5.2.
Schneider, J.-P. 'Boéthos de Sidon', *Dictionnaire des Philosophes Antiques*, vol. 2, Paris 1994, 126-30.
Sharples, R.W. (1) 'Alexander of Aphrodisias, *On Time*', *Phronesis* 27, 1982, 58-81.
—— (2) 'Alexander of Aphrodisias: scholasticism and innovation', *Aufstieg und Niedergang der Römischen Welt* II.36.2, 1987, 1176-1243.
—— (3) (trans.), *Alexander of Aphrodisias: Quaestiones 1.1-2.15*, London and Ithaca NY 1992. ACA
—— (4) 'Eudemus' *Physics*: Change, Place and Time', in W.W. Fortenbaugh and I. Bodnár (eds), *Eudemus of Rhodes*, New Brunswick 2002, 107-26. *RUSCH* 11
Shorey, P. 'Emendations of Themistius' Paraphrase of Aristotle's *Physics*', *Classical Philology* 3, 1908, 447-9.
Sicherl, M. *Die griechishen Erstausgaben des Vettore Trincavelli*, Paderborn 1993. *Studien zur Geschichte und Kultur des Altertums*, N.F., I. Reihe, 5. Band.
Simplic. *in Phys.* = H. Diels (ed.), *CAG* 9-10, Berlin 1882, 1895.
Sorabji, R. (1) *Time, Creation and the Continuum: Theories in Antiquity and the Early Middle Ages*, London 1983.
—— (2) 'General Introduction', 1-17 in C. Wildberg (trans.), *Philoponus: Against Aristotle on the Eternity of the World*, London and Ithaca NY 1986. ACA.
—— (3) *Matter, Space, and Motion: Theories in Antiquity and their Sequel*, London 1988.
—— (4) (ed.), *Aristotle Transformed: The Ancient Commentators and their Influence*, London 1990.
Spengel, L., *Themistii Paraphrases Aristotelis Librorum quae Supersunt*, Leipzig 1866; vol. 1, 253-342 = Themist. *in Phys.* 4.1.
SVF = H. von Arnim (ed.), *Stoicorum Veterum Fragmenta*, 3 vols, Leipzig 1903-5; repr. Stuttgart 1964.
Thorp, J. 'Aristotle's *Horror Vacui*', *Canadian Journal of Philosophy* 20, 1990, 149-66.
Todd, R.B. (1) 'Alexander of Aphrodisias and the Alexandrian *Quaestiones 2.12*', *Philologus* 116, 1972, 293-305.
—— (2) 'The Stoic common notions', *Symbolae Osloenses* 48, 1973, 47-75.
—— (3) *Alexander of Aphrodisias on Stoic Physics*, Leiden 1976. *Philosophia Antiqua* 28.
—— (4) (ed.), *Cleomedes: Caelestia*, Leipzig 1990.
—— (5) 'The alleged Stoic distinction between *holon* and *pan*', *Liverpool Classical Monthly* 16, 1991, 138-40.
—— (6) (trans.), *Themistius: On Aristotle on the Soul*, London and Ithaca NY 1996.
—— (7) 'An inventory of the Greek manuscripts of Themistius' Aristotelian commentaries', *Byzantion* 67, 1997, 268-76.

——— (8) 'Themistius', *Catalogus Translationum et Commentariorum* 8, 2003, 56-103.
Urmson, J.O. (1) (trans.), *Simplicius: On Aristotle's Physics 5.1-5, 10-14*, London 1992. *ACA*.
——— (2) (trans.), *Simplicius: Corollaries on Place and Time*, London and Ithaca NY 1992. *ACA*.
——— (3) (trans.), *Simplicius: On Aristotle on the Void*, London and Ithaca NY 1994 (= 159-267 in P. Lettinck [trans.], *Philoponus: On Aristotle's Physics 5-8*). *ACA*.
Usener, H. (ed.), *Epicurea*, Leipzig 1887; repr. Stuttgart 1966.
Wagner, H. (trans.), *Aristoteles: Physikvorlesung*, Berlin 1967.
Waterfield, R. (trans.), *Aristotle: Physics: Introduction and Notes by David Bostock*. Oxford 1996.
Wehrli, F. (ed.), *Die Schule des Aristoteles: I, Eudemos von Rhodos*, ed. 2, Basel 1969.
——— (ed.), *Die Schule des Aristotles: V, Straton von Lampsakos*, ed. 2, Basel 1969.
White, M.J. (1), 'Can unequal quantities of stuffs be totally blended?' *History of Philosophy Quarterly* 3, 1986, 379-89.
——— (2) 'Concepts of space in Greek thought' (review of Algra [2]), *Apeiron* 29, 1996, 183-98.
Wolff, M. 'Philoponus and the rise of preclassical dynamics', in R. Sorabji (ed.), *Philoponus and the Rejection of Aristotelian Science*, London and Ithaca NY 1987, 84-120.
Zonta, M. '*Hebraica Veritas*: Temistio, *Parafrasi* del *De Coelo*: Tradizione e Critica del Testo,' *Athenaeum* 82, 1994, 403-28.

Appendix: The Commentators*
Richard Sorabji

The 15,000 pages of the Ancient Greek Commentaries on Aristotle are the largest corpus of Ancient Greek philosophy that has not been translated into English or other European languages. The standard edition (*Commentaria in Aristotelem Graeca*, or *CAG*) was produced by Hermann Diels as general editor under the auspices of the Prussian Academy in Berlin. Arrangements have been made to translate at least a large proportion of this corpus, along with some other Greek and Latin commentaries not included in the Berlin edition, and some closely related non-commentary works by the commentators.

The works are not just commentaries on Aristotle, although they are invaluable in that capacity too. One of the ways of doing philosophy between A.D. 200 and 600, when the most important items were produced, was by writing commentaries. The works therefore represent the thought of the Peripatetic and Neoplatonist schools, as well as expounding Aristotle. Furthermore, they embed fragments from all periods of Ancient Greek philosophical thought: this is how many of the Presocratic fragments were assembled, for example. Thus they provide a panorama of every period of Ancient Greek philosophy.

The philosophy of the period from A.D. 200 to 600 has not yet been intensively explored by philosophers in English-speaking countries, yet it is full of interest for physics, metaphysics, logic, psychology, ethics and religion. The contrast with the study of the Presocratics is striking. Initially the incomplete Presocratic fragments might well have seemed less promising, but their interest is now widely known, thanks to the philological and philosophical effort that has been concentrated upon them. The incomparably vaster corpus which preserved so many of those fragments offers at least as much interest, but is still relatively little known.

The commentaries represent a missing link in the history of philosophy: the Latin-speaking Middle Ages obtained their knowledge of Aristotle at least partly through the medium of the commentaries. Without an appreciation of this, mediaeval interpretations of Aristotle will not be understood. Again, the ancient commentaries are the unsuspected source of ideas which have been thought, wrongly, to originate in the later mediaeval period. It has been supposed, for example, that Bonaventure in the thirteenth century invented the ingenious arguments based on the concept of infinity which attempt to prove the Christian view that the universe had a beginning. In fact, Bonaventure is merely repeating arguments devised

---

* Reprinted from the Editor's General Introduction to the series in Christian Wildberg, *Philoponus Against Aristotle on the Eternity of the World*, London and Ithaca, N.Y., 1987.

Appendix: The Commentators

by the commentator Philoponus 700 years earlier and preserved in the meantime by the Arabs. Bonaventure even uses Philoponus' original examples. Again, the introduction of impetus theory into dynamics, which has been called a scientific revolution, has been held to be an independent invention of the Latin West, even if it was earlier discovered by the Arabs or their predecessors. But recent work has traced a plausible route by which it could have passed from Philoponus, via the Arabs, to the West.

The new availability of the commentaries in the sixteenth century, thanks to printing and to fresh Latin translations, helped to fuel the Renaissance break from Aristotelian science. For the commentators record not only Aristotle's theories, but also rival ones, while Philoponus as a Christian devises rival theories of his own and accordingly is mentioned in Galileo's early works more frequently than Plato.[1]

It is not only for their philosophy that the works are of interest. Historians will find information about the history of schools, their methods of teaching and writing and the practices of an oral tradition.[2] Linguists will find the indexes and translations an aid for studying the development of word meanings, almost wholly uncharted in Liddell and Scott's *Lexicon*, and for checking shifts in grammatical usage.

Given the wide range of interests to which the volumes will appeal, the aim is to produce readable translations, and to avoid so far as possible presupposing any knowledge of Greek. Notes will explain points of meaning, give cross-references to other works, and suggest alternative interpretations of the text where the translator does not have a clear preference. The introduction to each volume will include an explanation why the work was chosen for translation: none will be chosen simply because it is there. Two of the Greek texts are currently being re-edited – those of Simplicius *in Physica* and *in de Caelo* – and new readings will be exploited by

---

1. See Fritz Zimmermann, 'Philoponus' impetus theory in the Arabic tradition'; Charles Schmitt, 'Philoponus' commentary on Aristotle's *Physics* in the sixteenth century', and Richard Sorabji, 'John Philoponus', in Richard Sorabji (ed.), *Philoponus and the Rejection of Aristotelian Science* (London and Ithaca, N.Y. 1987).

2. See e.g. Karl Praechter, 'Die griechischen Aristoteleskommentare', *Byzantinische Zeitschrift* 18 (1909), 516-38 (translated into English in R. Sorabji (ed.), *Aristotle Transformed: the ancient commentators and their influence* (London and Ithaca, N.Y. 1990); M. Plezia, *de Commentariis Isagogicis* (Cracow 1947); M. Richard, '*Apo Phônês*', Byzantion 20 (1950), 191-222; É. Evrard, *L'Ecole d'Olympiodore et la composition du commentaire à la physique de Jean Philopon*, Diss. (Liège 1957); L.G. Westerink, *Anonymous Prolegomena to Platonic Philosophy* (Amsterdam 1962) (new revised edition, translated into French, Collection Budé; part of the revised introduction, in English, is included in *Aristotle Transformed*); A.-J. Festugière, 'Modes de composition des commentaires de Proclus', *Museum Helveticum* 20 (1963), 77-100, repr. in his *Études* (1971), 551-74; P. Hadot, 'Les divisions des parties de la philosophie dans l'antiquité', *Museum Helveticum* 36 (1979), 201-23; I. Hadot, 'La division néoplatonicienne des écrits d'Aristote', in J. Wiesner (ed.), *Aristoteles Werk und Wirkung* (Paul Moraux gewidmet), vol. 2 (Berlin 1986); I. Hadot, 'Les introductions aux commentaires exégétiques chez les auteurs néoplatoniciens et les auteurs chrétiens', in M. Tardieu (ed.), *Les règles de l'interprétation* (Paris 1987), 99-119. These topics are treated, and a bibliography supplied, in *Aristotle Transformed*.

translators as they become available. Each volume will also contain a list of proposed emendations to the standard text. Indexes will be of more uniform extent as between volumes than is the case with the Berlin edition, and there will be at least three of them: an English-Greek glossary, a Greek-English index, and a subject index.

The commentaries fall into three main groups. The first group is by authors in the Aristotelian tradition up to the fourth century A.D. This includes the earliest extant commentary, that by Aspasius in the first half of the second century A.D. on the *Nicomachean Ethics*. The anonymous commentary on Books 2, 3, 4 and 5 of the *Nicomachean Ethics*, in *CAG* vol. 20, is derived from Adrastus, a generation later.[3] The commentaries by Alexander of Aphrodisias (appointed to his chair between A.D. 198 and 209) represent the fullest flowering of the Aristotelian tradition. To his successors Alexander was The Commentator *par excellence*. To give but one example (not from a commentary) of his skill at defending and elaborating Aristotle's views, one might refer to his defence of Aristotle's claim that space is finite against the objection that an edge of space is conceptually problematic.[4] Themistius (*fl.* late 340s to 384 or 385) saw himself as the inventor of paraphrase, wrongly thinking that the job of commentary was completed.[5] In fact, the Neoplatonists were to introduce new dimensions into commentary. Themistius' own relation to the Neoplatonist as opposed to the Aristotelian tradition is a matter of controversy,[6] but it would be agreed that his commentaries show far less bias than the full-blown Neoplatonist ones. They are also far more informative than the designation 'paraphrase' might suggest, and it has been estimated that Philoponus' *Physics* commentary draws silently on Themistius six hundred times.[7] The pseudo-Alexandrian commentary on *Metaphysics* 6-14, of unknown authorship, has been

---

3. Anthony Kenny, *The Aristotelian Ethics* (Oxford 1978), 37, n.3: Paul Moraux, *Der Aristotelismus bei den Griechen*, vol. 2 (Berlin 1984), 323-30.

4. Alexander, *Quaestiones* 3.12, discussed in my *Matter, Space and Motion* (London and Ithaca, N.Y. 1988). For Alexander see R.W. Sharples, 'Alexander of Aphrodisias: scholasticism and innovation', in W. Haase (ed.), *Aufstieg und Niedergang der römischen Welt*, part 2 *Principat*, vol. 36.2, *Philosophie und Wissenschaften* (1987).

5. Themistius *in An. Post.* 1,2-12. See H.J. Blumenthal, 'Photius on Themistius (Cod. 74): did Themistius write commentaries on Aristotle?', *Hermes* 107 (1979), 168-82.

6. For different views, see H.J. Blumenthal, 'Themistius, the last Peripatetic commentator on Aristotle?', in Glen W. Bowersock, Walter Burkert, Michael C.J. Putnam, *Arktouros*, Hellenic Studies Presented to Bernard M.W. Knox (Berlin and N.Y., 1979), 391-400; E.P. Mahoney, 'Themistius and the agent intellect in James of Viterbo and other thirteenth-century philosophers: (Saint Thomas Aquinas, Siger of Brabant and Henry Bate)', *Augustiniana* 23 (1973), 422-67, at 428-31; id., 'Neoplatonism, the Greek commentators and Renaissance Aristotelianism', in D.J. O'Meara (ed.), *Neoplatonism and Christian Thought* (Albany N.Y. 1982), 169-77 and 264-82, esp. n. 1, 264-6; Robert Todd, introduction to translation of Themistius *in DA* 3.4-8, in *Two Greek Aristotelian Commentators on the Intellect*, trans. Frederick M. Schroeder and Robert B. Todd (Toronto 1990).

7. H. Vitelli, *CAG* 17, p. 992, s.v. Themistius.

placed by some in the same group of commentaries as being earlier than the fifth century.[8]

By far the largest group of extant commentaries is that of the Neoplatonists up to the sixth century A.D. Nearly all the major Neoplatonists, apart from Plotinus (the founder of Neoplatonism), wrote commentaries on Aristotle, although those of Iamblichus (c. 250–c. 325) survive only in fragments, and those of three Athenians, Plutarchus (died 432), his pupil Proclus (410–485) and the Athenian Damascius (c. 462–after 538), are lost.[9] As a result of these losses, most of the extant Neoplatonist commentaries come from the late fifth and the sixth centuries and a good proportion from Alexandria. There are commentaries by Plotinus' disciple and editor Porphyry (232–309), by Iamblichus' pupil Dexippus (c. 330), by Proclus' teacher Syrianus (died c. 437), by Proclus' pupil Ammonius (435/445–517/526), by Ammonius' three pupils Philoponus (c. 490 to 570s), Simplicius (wrote after 532, probably after 538) and Asclepius (sixth century), by Ammonius' next but one successor Olympiodorus (495/505–after 565), by Elias (fl. 541?), by David (second half of the sixth century, or beginning of the seventh) and by Stephanus (took the chair in Constantinople c. 610). Further, a commentary on the *Nicomachean Ethics* has been ascribed to Heliodorus of Prusa, an unknown pre-fourteenth-century figure, and there is a commentary by Simplicius' colleague Priscian of Lydia on Aristotle's successor Theophrastus. Of these commentators some of the last were Christians (Philoponus, Elias, David and Stephanus), but they were Christians writing in the Neoplatonist tradition, as was also Boethius who produced a number of commentaries in Latin before his death in 525 or 526.

The third group comes from a much later period in Byzantium. The Berlin edition includes only three out of more than a dozen commentators described in Hunger's *Byzantinisches Handbuch*.[10] The two most important are Eustratius (1050/1060–c.1120), and Michael of Ephesus. It has been suggested that these two belong to a circle organised by the princess

---

8. The similarities to Syrianus (died c. 437) have suggested to some that it predates Syrianus (most recently Leonardo Tarán, review of Paul Moraux, *Der Aristotelismus*, vol.1 in *Gnomon* 46 (1981), 721-50 at 750), to others that it draws on him (most recently P. Thillet, in the Budé edition of Alexander *de Fato*, p. lvii). Praechter ascribed it to Michael of Ephesus (eleventh or twelfth century), in his review of *CAG* 22.2, in *Göttingische Gelehrte Anzeiger* 168 (1906), 861-907.

9. The Iamblichus fragments are collected in Greek by Bent Dalsgaard Larsen, *Jamblique de Chalcis, Exégète et Philosophe* (Aarhus 1972), vol. 2. Most are taken from Simplicius, and will accordingly be translated in due course. The evidence on Damascius' commentaries is given in L.G. Westerink, *The Greek Commentaries on Plato's Phaedo*, vol. 2, Damascius (Amsterdam 1977), 11-12; on Proclus' in L.G. Westerink, *Anonymous Prolegomena to Platonic Philosophy* (Amsterdam 1962), xii, n. 22; on Plutarchus' in H.M. Blumenthal, 'Neoplatonic elements in the de Anima commentaries', *Phronesis* 21 (1976), 75.

10. Herbert Hunger, *Die hochsprachliche profane Literatur der Byzantiner*, vol. 1 (= *Byzantinisches Handbuch*, part 5, vol. 1) (Munich 1978), 25-41. See also B.N. Tatakis, *La Philosophie Byzantine* (Paris 1949).

Anna Comnena in the twelfth century, and accordingly the completion of Michael's commentaries has been redated from 1040 to 1138.[11] His commentaries include areas where gaps had been left. Not all of these gap-fillers are extant, but we have commentaries on the neglected biological works, on the *Sophistici Elenchi*, and a small fragment of one on the *Politics*. The lost *Rhetoric* commentary had a few antecedents, but the *Rhetoric* too had been comparatively neglected. Another product of this period may have been the composite commentary on the *Nicomachean Ethics* (*CAG* 20) by various hands, including Eustratius and Michael, along with some earlier commentators, and an improvisation for Book 7. Whereas Michael follows Alexander and the conventional Aristotelian tradition, Eustratius' commentary introduces Platonist, Christian and anti-Islamic elements.[12]

The composite commentary was to be translated into Latin in the next century by Robert Grosseteste in England. But Latin translations of various logical commentaries were made from the Greek still earlier by James of Venice (*fl. c.* 1130), a contemporary of Michael of Ephesus, who may have known him in Constantinople. And later in that century other commentaries and works by commentators were being translated from Arabic versions by Gerard of Cremona (died 1187).[13] So the twelfth century resumed the transmission which had been interrupted at Boethius' death in the sixth century.

The Neoplatonist commentaries of the main group were initiated by Porphyry. His master Plotinus had discussed Aristotle, but in a very independent way, devoting three whole treatises (*Enneads* 6.1-3) to attacking Aristotle's classification of the things in the universe into categories. These categories took no account of Plato's world of Ideas, were inferior to Plato's classifications in the *Sophist* and could anyhow be collapsed, some

---

11. R. Browning, 'An unpublished funeral oration on Anna Comnena', *Proceedings of the Cambridge Philological Society* n.s. 8 (1962), 1-12, esp. 6-7.

12. R. Browning, op. cit. H.D.P. Mercken, *The Greek Commentaries of the Nicomachean Ethics of Aristotle in the Latin Translation of Grosseteste*, Corpus Latinum Commentariorum in Aristotelem Graecorum VI 1 (Leiden 1973), ch. 1, 'The compilation of Greek commentaries on Aristotle's Nicomachean Ethics'. Sten Ebbesen, 'Anonymi Aurelianensis I Commentarium in *Sophisticos Elenchos*', *Cahiers de l'Institut Moyen Age Grecque et Latin* 34 (1979), 'Boethius, Jacobus Veneticus, Michael Ephesius and "Alexander" ', pp. v-xiii; id., *Commentators and Commentaries on Aristotle's Sophistici Elenchi*, 3 parts, Corpus Latinum Commentariorum in Aristotelem Graecorum, vol. 7 (Leiden 1981); A. Preus, *Aristotle and Michael of Ephesus on the Movement and Progression of Animals* (Hildesheim 1981), introduction.

13. For Grosseteste, see Mercken as in n. 12. For James of Venice, see Ebbesen as in n. 12, and L. Minio-Paluello, 'Jacobus Veneticus Grecus', *Traditio* 8 (1952), 265-304; id., 'Giacomo Veneto e l'Aristotelismo Latino', in Pertusi (ed.), *Venezia e l'Oriente fra tardo Medioevo e Rinascimento* (Florence 1966), 53-74, both reprinted in his *Opuscula* (1972). For Gerard of Cremona, see M. Steinschneider, *Die europäischen Übersetzungen aus dem arabischen bis Mitte des 17. Jahrhunderts* (repr. Graz 1956); E. Gilson, *History of Christian Philosophy in the Middle Ages* (London 1955), 235-6 and more generally 181-246. For the translators in general, see Bernard G. Dod, 'Aristoteles Latinus', in N. Kretzmann, A. Kenny, J. Pinborg (eds), *The Cambridge History of Latin Medieval Philosophy* (Cambridge 1982).

of them into others. Porphyry replied that Aristotle's categories could apply perfectly well to the world of intelligibles and he took them as in general defensible.[14] He wrote two commentaries on the *Categories*, one lost, and an introduction to it, the *Isagôgê*, as well as commentaries, now lost, on a number of other Aristotelian works. This proved decisive in making Aristotle a necessary subject for Neoplatonist lectures and commentary. Proclus, who was an exceptionally quick student, is said to have taken two years over his Aristotle studies, which were called the Lesser Mysteries, and which preceded the Greater Mysteries of Plato.[15] By the time of Ammonius, the commentaries reflect a teaching curriculum which begins with Porphyry's *Isagôgê* and Aristotle's *Categories*, and is explicitly said to have as its final goal a (mystical) ascent to the supreme Neoplatonist deity, the One.[16] The curriculum would have progressed from Aristotle to Plato, and would have culminated in Plato's *Timaeus* and *Parmenides*. The latter was read as being about the One, and both works were established in this place in the curriculum at least by the time of Iamblichus, if not earlier.[17]

Before Porphyry, it had been undecided how far a Platonist should accept Aristotle's scheme of categories. But now the proposition began to gain force that there was a harmony between Plato and Aristotle on most things.[18] Not for the only time in the history of philosophy, a perfectly crazy proposition proved philosophically fruitful. The views of Plato and of Aristotle had both to be transmuted into a new Neoplatonist philosophy in order to exhibit the supposed harmony. Iamblichus denied that Aristotle contradicted Plato on the theory of Ideas.[19] This was too much for Syrianus and his pupil Proclus. While accepting harmony in many areas,[20] they could see that there was disagreement on this issue and also on the issue of whether God was causally responsible for the existence of the ordered

---

14. See P. Hadot, 'L'harmonie des philosophies de Plotin et d'Aristote selon Porphyre dans le commentaire de Dexippe sur les Catégories', in *Plotino e il neoplatonismo in Oriente e in Occidente* (Rome 1974), 31-47; A.C. Lloyd, 'Neoplatonic logic and Aristotelian logic', *Phronesis* 1 (1955-6), 58-79 and 146-60.
15. Marinus, *Life of Proclus* ch. 13, 157,41 (Boissonade).
16. The introductions to the *Isagôgê* by Ammonius, Elias and David, and to the *Categories* by Ammonius, Simplicius, Philoponus, Olympiodorus and Elias are discussed by L.G. Westerink, *Anonymous Prolegomena* and I. Hadot, 'Les Introductions', see n. 2 above.
17. Proclus in *Alcibiadem 1* p. 11 (Creuzer); Westerink, *Anonymous Prolegomena*, ch. 26, 12f. For the Neoplatonist curriculum see Westerink, Festugière, P. Hadot and I. Hadot in n. 2.
18. See e.g. P. Hadot (1974), as in n. 14 above; H.J. Blumenthal, 'Neoplatonic elements in the de Anima commentaries', *Phronesis* 21 (1976), 64-87; H.A. Davidson, 'The principle that a finite body can contain only finite power', in S. Stein and R. Loewe (eds), *Studies in Jewish Religious and Intellectual History presented to A. Altmann* (Alabama 1979), 75-92; Carlos Steel, 'Proclus et Aristotle', Proceedings of the Congrès Proclus held in Paris 1985, J. Pépin and H.D. Saffrey (eds), *Proclus, lecteur et interprète des anciens* (Paris 1987), 213-25; Koenraad Verrycken, *God en Wereld in de Wijsbegeerte van Ioannes Philoponus*, Ph.D. Diss. (Louvain 1985).
19. Iamblichus ap. Elian in *Cat.* 123,1-3.
20. Syrianus in *Metaph.* 80,4-7; Proclus in *Tim.* 1.6,21-7,16.

physical cosmos, which Aristotle denied. But even on these issues, Proclus' pupil Ammonius was to claim harmony, and, though the debate was not clear cut,[21] his claim was on the whole to prevail. Aristotle, he maintained, accepted Plato's Ideas,[22] at least in the form of principles (*logoi*) in the divine intellect, and these principles were in turn causally responsible for the beginningless existence of the physical universe. Ammonius wrote a whole book to show that Aristotle's God was thus an efficent cause, and though the book is lost, some of its principal arguments are preserved by Simplicius.[23] This tradition helped to make it possible for Aquinas to claim Aristotle's God as a Creator, albeit not in the sense of giving the universe a beginning, but in the sense of being causally responsible for its beginningless existence.[24] Thus what started as a desire to harmonise Aristotle with Plato finished by making Aristotle safe for Christianity. In Simplicius, who goes further than anyone,[25] it is a formally stated duty of the commentator to display the harmony of Plato and Aristotle in most things.[26] Philoponus, who with his independent mind had thought better of his earlier belief in harmony, is castigated by Simplicius for neglecting this duty.[27]

The idea of harmony was extended beyond Plato and Aristotle to Plato and the Presocratics. Plato's pupils Speusippus and Xenocrates saw Plato as being in the Pythagorean tradition.[28] From the third to first centuries B.C., pseudo-Pythagorean writings present Platonic and Aristotelian doctrines as if they were the ideas of Pythagoras and his pupils,[29] and these forgeries were later taken by the Neoplatonists as genuine. Plotinus saw the Presocratics as precursors of his own views,[30] but Iamblichus went far beyond him by writing ten volumes on Pythagorean philosophy.[31] Thereafter Proclus sought to unify the whole of Greek

21. Asclepius sometimes accepts Syranius' interpretation (*in Metaph.* 433,9-436,6); which is, however, qualified, since Syrianus thinks Aristotle is realy committed willy-nilly to much of Plato's view (*in Metaph.* 117,25-118,11; ap. Asclepium *in Metaph.* 433,16; 450,22); Philoponus repents of his early claim that Plato is not the target of Aristotle's attack, and accepts that Plato is rightly attacked for treating ideas as independent entities outside the divine Intellect (*in DA* 37,18-31; *in Phys.* 225,4-226,11; *contra Procl.* 26,24-32,13; *in An. Post.* 242,14-243,25).

22. Asclepius *in Metaph.* from the voice of (i.e. from the lectures of) Ammonius 69,17-21; 71,28; cf. Zacharias *Ammonius, Patrologia Graeca* vol. 85 col. 952 (Colonna).

23. Simplicius *in Phys.* 1361,11-1363,12. See H.A. Davidson; Carlos Steel; Koenraad Verrycken in n. 18 above.

24. See Richard Sorabji, *Matter, Space and Motion* (London and Ithaca, N.Y. 1988), ch. 15.

25. See e.g. H.J. Blumenthal in n. 18 above.

26. Simplicius *in Cat.* 7,23-32.

27. Simplicius *in Cael.* 84,11-14; 159,2-9. On Philoponus' *volte face* see n. 21 above.

28. See e.g. Walter Burkert, *Weisheit und Wissenschaft* (Nürnberg 1962), translated as *Lore and Science in Ancient Pythagoreanism* (Cambridge Mass. 1972), 83-96.

29. See Holger Thesleff, *An Introduction to the Pythagorean Writings of the Hellenistic Period* (Åbo 1961); Thomas Alexander Szlezák, *Pseudo-Archytas über die Kategorien*, Peripatoi vol. 4 (Berlin and New York 1972).

30. Plotinus e.g. 4.8.1; 5.1.8 (10-27); 5.1.9.

31. See Dominic O'Meara, *Pythagoras Revived: Mathematics and Philosophy in Late Antiquity* (Oxford 1989).

Appendix: The Commentators

philosophy by presenting it as a continuous clarification of divine revelation[32] and Simplicius argued for the same general unity in order to rebut Christian charges of contradictions in pagan philosophy.[33]

Later Neoplatonist commentaries tend to reflect their origin in a teaching curriculum:[34] from the time of Philoponus, the discussion is often divided up into lectures, which are subdivided into studies of doctrine and of text. A general account of Aristotle's philosophy is prefixed to the *Categories* commentaries and divided, according to a formula of Proclus,[35] into ten questions. It is here that commentators explain the eventual purpose of studying Aristotle (ascent to the One) and state (if they do) the requirement of displaying the harmony of Plato and Aristotle. After the ten-point introduction to Aristotle, the *Categories* is given a six-point introduction, whose antecedents go back earlier than Neoplatonism, and which requires the commentator to find a unitary theme or scope (*skopos*) for the treatise. The arrangements for late commentaries on Plato are similar. Since the Plato commentaries form part of a single curriculum they should be studied alongside those on Aristotle. Here the situation is easier, not only because the extant corpus is very much smaller, but also because it has been comparatively well served by French and English translators.[36]

Given the theological motive of the curriculum and the pressure to harmonise Plato with Aristotle, it can be seen how these commentaries are a major source for Neoplatonist ideas. This in turn means that it is not safe to extract from them the fragments of the Presocratics, or of other authors, without making allowance for the Neoplatonist background against which the fragments were originally selected for discussion. For different reasons, analogous warnings apply to fragments preserved by the pre- Neoplatonist commentator Alexander.[37] It will be another advantage of the present translations that they will make it easier to check the distorting effect of a commentator's background.

Although the Neoplatonist commentators conflate the views of Aristotle

---

32. See Christian Guérard, 'Parménide d'Elée selon les Néoplatoniciens', in P. Aubenque (ed.), *Etudes sur Parménide*, vol. 2 (Paris 1987).

33. Simplicius *in Phys.* 28,32-29,5; 640,12-18. Such thinkers as Epicurus and the Sceptics, however, were not subject to harmonisation.

34. See the literature in n. 2 above.

35. ap. Elian *in Cat.* 107,24-6.

36. English: Calcidius *in Tim.* (parts by van Winden; den Boeft); Iamblichus fragments (Dillon); Proclus *in Tim.* (Thomas Taylor); Proclus *in Parm.* (Dillon); Proclus *in Parm.*, end of 7th book, from the Latin (Klibansky, Labowsky, Anscombe); Proclus *in Alcib. 1* (O'Neill); Olympiodorus and Damascius *in Phaedonem* (Westerink); Damascius *in Philebum* (Westerink); *Anonymous Prolegomena to Platonic Philosophy* (Westerink). See also extracts in Thomas Taylor, *The Works of Plato*, 5 vols. (1804). French: Proclus *in Tim.* and *in Rempublicam* (Festugière); *in Parm.* (Chaignet); Anon. *in Parm* (P. Hadot); Damascius *in Parm.* (Chaignet).

37. For Alexander's treatment of the Stoics, see Robert B. Todd, *Alexander of Aphrodisias on Stoic Physics* (Leiden 1976), 24-9.

with those of Neoplatonism, Philoponus alludes to a certain convention when he quotes Plutarchus expressing disapproval of Alexander for expounding his own philosophical doctrines in a commentary on Aristotle.[38] But this does not stop Philoponus from later inserting into his own commentaries on the *Physics* and *Meteorology* his arguments in favour of the Christian view of Creation. Of course, the commentators also wrote independent works of their own, in which their views are expressed independently of the exegesis of Aristotle. Some of these independent works will be included in the present series of translations.

The distorting Neoplatonist context does not prevent the commentaries from being incomparable guides to Aristotle. The introductions to Aristotle's philosophy insist that commentators must have a minutely detailed knowledge of the entire Aristotelian corpus, and this they certainly have. Commentators are also enjoined neither to accept nor reject what Aristotle says too readily, but to consider it in depth and without partiality. The commentaries draw one's attention to hundreds of phrases, sentences and ideas in Aristotle, which one could easily have passed over, however often one read him. The scholar who makes the right allowance for the distorting context will learn far more about Aristotle than he would be likely to on his own.

The relations of Neoplatonist commentators to the Christians were subtle. Porphyry wrote a treatise explicitly against the Christians in 15 books, but an order to burn it was issued in 448, and later Neoplatonists were more circumspect. Among the last commentators in the main group, we have noted several Christians. Of these the most important were Boethius and Philoponus. It was Boethius' programme to transmit Greek learning to Latin-speakers. By the time of his premature death by execution, he had provided Latin translations of Aristotle's logical works, together with commentaries in Latin but in the Neoplatonist style on Porphyry's *Isagôgê* and on Aristotle's *Categories* and *de Interpretatione*, and interpretations of the *Prior* and *Posterior Analytics*, *Topics* and *Sophistici Elenchi*. The interruption of his work meant that knowledge of Aristotle among Latin-speakers was confined for many centuries to the logical works. Philoponus is important both for his proofs of the Creation and for his progressive replacement of Aristotelian science with rival theories, which were taken up at first by the Arabs and came fully into their own in the West only in the sixteenth century.

Recent work has rejected the idea that in Alexandria the Neoplatonists compromised with Christian monotheism by collapsing the distinction between their two highest deities, the One and the Intellect. Simplicius (who left Alexandria for Athens) and the Alexandrians Ammonius and

---

38. Philoponus *in DA* 21,20-3.

Asclepius appear to have acknowledged their beliefs quite openly, as later did the Alexandrian Olympiodorus, despite the presence of Christian students in their classes.[39]

The teaching of Simplicius in Athens and that of the whole pagan Neoplatonist school there was stopped by the Christian Emperor Justinian in 529. This was the very year in which the Christian Philoponus in Alexandria issued his proofs of Creation against the earlier Athenian Neoplatonist Proclus. Archaeological evidence has been offered that, after their temporary stay in Ctesiphon (in present-day Iraq), the Athenian Neoplatonists did not return to their house in Athens, and further evidence has been offered that Simplicius went to Harrān (Carrhae), in present-day Turkey near the Iraqi border.[40] Wherever he went, his commentaries are a treasurehouse of information about the preceding thousand years of Greek philosophy, information which he painstakingly recorded after the closure in Athens, and which would otherwise have been lost. He had every reason to feel bitter about Christianity, and in fact he sees it and Philoponus, its representative, as irreverent. They deny the divinity of the heavens and prefer the physical relics of dead martyrs.[41] His own commentaries by contrast culminate in devout prayers.

Two collections of articles by various hands have been published, to make the work of the commentators better known. The first is devoted to Philoponus;[42] the second is about the commentators in general, and goes into greater detail on some of the issues briefly mentioned here.[43]

---

39. For Simplicius, see I. Hadot, *Le Problème du Néoplatonisme Alexandrin: Hiéroclès et Simplicius* (Paris 1978); for Ammonius and Asclepius, Koenraad Verrycken, *God en wereld in de Wijsbegeerte van Ioannes Philoponus*, Ph.D. Diss. (Louvain 1985); for Olympiodorus, L.G. Westerink, *Anonymous Prolegomena to Platonic Philosophy* (Amsterdam 1962).

40. Alison Frantz, 'Pagan philosophers in Christian Athens', *Proceedings of the American Philosophical Society* 119 (1975), 29-38; M. Tardieu, 'Témoins orientaux du *Premier Alcibiade* à Harrān et à Nag 'Hammādi', *Journal Asiatique* 274 (1986); id., 'Les calendriers en usage à Harrān d'après les sources arabes et le commentaire de Simplicius à la *Physique* d'Aristote', in I. Hadot (ed.), *Simplicius, sa vie, son oeuvre, sa survie* (Berlin 1987), 40-57; id., *Coutumes nautiques mésopotamiennes chez Simplicius*, in preparation. The opposing view that Simplicius returned to Athens is most fully argued by Alan Cameron, 'The last days of the Academy at Athens', *Proceedings of the Cambridge Philological Society* 195, n.s. 15 (1969), 7-29. P. Foulkes, 'Where was Simplicius?', *JHS* 112 (1992), 143. R. Thiel, 'Simplikios und das Ende der neuplatonischen Schule in Athen', Akademie der Wissenschaften und der Literatur Mainz: *Abhandlungen der geistes- und sozialwissenschaftlichen Klasse*, no. 8, 1999.

41. Simplicius *in Cael.* 26,4-7; 70,16-18; 90,1-18; 370,29-371,4. See on his whole attitude Philippe Hoffmann, 'Simplicius' polemics', in Richard Sorabji (ed.), *Philoponus and the Rejection of Aristotelian Science* (London and Ithaca, N.Y. 1987).

42. Richard Sorabji (ed.), *Philoponus and the Rejection of Aristotelian Science* (London and Ithaca, N.Y. 1987).

43. Richard Sorabji (ed.), *Aristotle Transformed: the ancient commentators and their influence* (London and Ithaca, N.Y. 1990). The lists of texts and previous translations of the commentaries included in Wildberg, *Philoponus Against Aristotle on the Eternity of the World* (pp. 12ff.) are not included here. The list of translations should be augmented by: F.L.S. Bridgman, Heliodorus (?) *in Ethica Nicomachea*, London 1807.

I am grateful for comments to Henry Blumenthal, Victor Caston, I. Hadot, Paul Mercken, Alain Segonds, Robert Sharples, Robert Todd, L.G. Westerink and Christian Wildberg.

*

The foregoing reprint of the editor's 1987 introduction shows how the subject has advanced in fifteen years. Many of the plans announced in the future tense are now realised, and in 2004 we shall publish an updated Sourcebook in three volumes, *The Philosophy of the Commentators, 200-600 AD: A Sourcebook in Translation*. Meanwhile the translator has kindly supplied some new bibliography, as follows.

To note 4, P. Moraux, *Der Aristotelismus bei den Griechen von Andronikos bis Alexander von Aphrodisias, Bd. III: Alexander von Aphrodisias*, edited by J. Wiesner with contributions in English by R.W. Sharples, Berlin, 2001 (*Peripatoi* 7/1).

To note 6, Robert B. Todd, 'Introduction' to *Themistius On Aristotle On the Soul*, London and Ithaca, NY, 1996.

To note 9, H.J. Blumenthal, *Aristotle and Neoplatonism in Late Antiquity: Interpretations of the* De anima, London and Ithaca, NY, 1996.

To note 10: B.N. Tatakis, *Byzantine Philosophy*, trans. into English by N. Moutafakis, Indianapolis, 2002.

To note 10: M. Share, *Arethas of Caesarea's Scholia on Porphyry's* Isagogê *and Aristotle's* Categories (Brussels, 1994) is the first volume of the projected *Commentaria in Aristotelem Byzantina*.

English-Greek Glossary

actual, in actuality: *energeiâi, kat'energeian*
affected, be: *paskhein*
affection: *pathos*
after: *husteron*
air: *aêr*
All, the (sc. the cosmos): *to pan*
alteration: *alloiôsis*
animal: *zôon*
argument: *logos*
assume: *hupolambanein, hupotithesthai*
assumption: *hupolêpsis, hupothesis*

be: *einai*
before: *proteron*
being: *to einai*
belief: *doxa*
believe: *dokein, hupolambanein, oiesthai*
bodies, natural: *phusika sômata*
bodies, uncompounded: *hapla sômata*
body: *sôma*
bound (v.): *horizein*
boundary: *horos*

carried, be: *pheresthai*
cease to be: *phtheiresthai*
ceasing to be: *phthora*
centre: *to meson*
change: *kinêsis*; in respect of place: *kinêsis kata topon*
circle: *kuklos*
clear: *dêlos, phaneros*
come to be, into being: *ginesthai*
coming into being: *genesis*
compact: *pakhus, puknos*
compacted, be: *pakhunesthai, puknousthai*
compacting: *puknôsis*
compressed, be: *pileisthai*
compression: *pilêsis*
conceive: *noein, ennoein*
conception: *epinoia, noêsis*
contact, be in: *haptesthai*

contain: *periekhein*
container: *to periekhon*
continuous: *sunekhês*
contracted, be: *sustellesthai*
contraction: *sustolê*
cosmos: *kosmos, ouranos*
count (v.): *arithmein*
culmination: *akmê*

(becoming) denser: *puknôsis*
(less) dense: *manos*
(more) dense: *puknos*
day: *hêmera*
decline (n.): *parakmê*
decrease: *phthisis*
define: *horizein, legein*
definition: *logos*
delimited, be: *peratousthai*
demarcate: *diorizein*
demonstrate: *deiknunai, epideiknunai*
destruction: *phthora*
difference: *diaphora*
diminution: *phthisis*
dispersed, be: *paraspeiresthai*
distinct feature: *diaphora*
distinguish: *diorizein*
divide: *diairein*
divisible: *diairetos*
doctrine: *doxa*
down(wards): *katô*

earth: *gê*
element: *stoikheion*
eliminate (through argument): *anairein*
end: *telos*
enter: *ginesthai en*
essence: *ousia, to ti ên einai*
examine: *exetazein*
exist: *einai, huparkhein*
exist (come to): *ginesthai*
existence (coming into): *genesis*
expand: *diakheisthai*
explain: *apodidonai*
extension: *diastêma*

extremity: *eskhaton*

fire: *pur*
first principle: *arkhê*
follow (logically): *akolouthein*
form: *eidos*

general, in: *holôs*
genus: *genos*

heavens: *ouranos*
heaviness: *barutês*
heavy: *barus*
hour: *hôra*
human being: *anthrôpos*

identify: *lambanein*

imagine: *phantazesthai*
impede: *empodizein, kôluein*
impossible: *adunatos, amêkhanos*
incidental, be: *sumbebêkenai*
incidentally: *kata sumbebêkos*
incorporeal: *asômatos*
increase/be increased: *auxanesthai, auxesthai*
increase: *auxêsis*
inquire into: *theôrein, zêtein*
inquiry: *skemma, theoria, zêtêsis*
inseparable: *akhôristos*
intelligible: *noêtos*
investigate: *episkeptesthai, skeptesthai*

kind: *eidos*
kindred: *sungenês*

light: *kouphos*
lightness: *kouphotês*
limit (n.): *peras*
limit (delimit) (v.): *peratoun*
line: *grammê*
locomotion: *phora (kata topon)*

magnitude: *megethos*
matter: *hulê*
measure (n.): *metron*
measure (v.): *metrein*
measure out (in units): *katametrein*
month: *mên*
motion: *phora*
move (sc. engage in locomotion): *kineisthai*
movement: *kinêsis*; forced: *biâi,*

biaios; natural: *kata phusin*;
contra-natural: *para phusin*
mutual replacement (engage in): *antimetastasis, antimethistasthai*

natural philosopher: *phusikos*
nature: *phusis*
nature, by: *phusei*
natural: *phusikos*
necessary: *anankaion*
necessarily: *anankaiôs, anankê esti*
notion: *ennoia*
notion, form a: *ennoein*
now, the: *to nun*
number (n.): *arithmos*

object (physical): *pragma*
obvious: *dêlos, phaneros*
occupy: *katekhein*

part: *meros, morion*
per se: *kath' hauto, kath' hauta*
perceptible: *aisthêtos*
perception: *aisthêsis*
persist: *anamenein, diamenein*
place: *topos*
plenum: *to plêres*
point: *stigmê*
posit: *tithenai, tithesthai*
position: *thesis*
possible: *dunatos*
pot: *kaddos*
potential, in potentiality: *dunamei, kata dunamin*
power: *dunamis*
primary: *prôtos*
problem, raise a: *aporein*
problem: *aporia*
proper: *oikeios*
proprietary: *idios*
rare: *manos*
rarefaction: *manôsis*
rarefied (become): *manousthai*
rarefied (becoming): *manôsis*
ratio: *logos*
receive: *dekhesthai*
recognise: *gnôrizein*
remain: *menein*
rest (n.): *êremia, monê*
rest, be at: *êremein*
revolution (celestial): *periphora*
revolution (sc. circular motion): *kuklophoria*

say: *legein*
separate (adj.): *khôristos*
separate (v.): *diakrinein, khôrizein*
signify: *sêmainein*
solve (a problem): *luein*
soul: *psukhê*
shape: *skhêma*
space: *khôra*
speak of: *legein*
speed: *takhos*
state: *legein*
structure, formal structure: *morphê*
subsist: *huphistasthai*
substance: *ousia*
substrate: *hupokeimenon*
successively different: *allo kai allo*

supply (answer; definition): *apodidonai*
surface: *epiphaneia*

tenuous: *leptos*
terminus: *peras, teleutê*

think: *noein, oiesthai*
time: *khronos*; in time: *en khronôi*
transformation: *metabolê*
transformed, be: *metaballein*

uncompounded bodies: *hapla sômata*
underlie: *hupokeisthai*
undifferentiated: *adiaphoros*
unit: *monas*
up(wards): *anô*

vessel: *angeion, skeuos*
void: *to kenon*; dispersed: *paresparmenon*; gross: *athroon*; intermingled: *enkekramenon*; separate: *kekhôrisomenon*
volume: *onkos*

water: *hudôr*
whole, the: *to holon*

year: *eniautos*

Greek-English Index

For a lexicographical guide to Aristotle's *Physics*, with a statistical analysis of usage by books, see B. Colin, *Aristote: Physica, Index verborum; Listes de fréquence*, Liège 1993. Its statistics are used where ratios are given for Aristotelian to Themistian usage. Asterisks (*) indicate loci that contain emendations.

abasanistôs, without examination, 163,8
adiairetos, undivided, 123,19; 151,3; 145,4
adiaphoros, undifferentiated, 122,23; 128,3.10
adunatos, impossible, 104,19.30; 107,1; 108,23; 110,5; 124,7; 125,13.28; 133,1; 134,31; 142,15
aisthanesthai, perceive, 132,24; 143,27; 144,16.18.30; 145,3.4.5; 160,28
aisthêsis, perception, 118,22; 125,19; 133,21; 142,29.30; 144,7.13.21; 147,7
aisthêtos, perceptible, 115,25; 123,6.7; 125,7-19; *to aisthêton*, object of perception, 160,27
aithêr, aether, 121,20(bis)
aitia, cause, explanation, 113,12; 124,18; 128,1; 129,19; 130,22; 132,8.14.27; 133,12; 136,25; 159,7.11.15
aitiasthai, identify as a cause, 123,18; 138,23; 139,18; 142,11
aition, cause, explanation, 111,17; 115,11; 118,16; 126,12-18; 127,29-33; 128,3.5.19; 136,16-26; 140,2.5; 145,22; 152,5; 155,17
akhôristos, inseparable, 107,2; 121,5; 128,14; 134,11; 137,2; 145,18; 145,15
akinêtos, unmoved, 117,27.29; 118,14.30; 119,2.5; unchanging, 144,27; 156,11.12
akolouthein, to follow (be implied by), to be a consequence (for someone), 102,8; 125,19; 128,27; 132,7; 145,12; 150,12.17; 153,24; 155,8; adhere to, 115,15
akolouthia, (logical) sequence, 102,3
akouein, hear (= attend to a text), 144,29
alêtheia, truth, 144,10
alêthês, true, 102,10.18; 105,16; 108,23; 112,23; 121,15; 123,19; 129,12; 132,5; 136,15; 160,11; *alethôs*, truly, 121,11; 133,16
alêtheuesthai, be truly stated, 105,27
alloiousthai, be altered, 137,30; 138,18; 161,6
alloiôsis, alteration, 126,19; 140,6; 152,12; 161,11.13.15.31; 162,7; 163,14
alogos, illogical, 143,1; *alogôs*, 163,27
ameibein, cause change, 119,19.22; 120,18; 126,21
amêkhanos, impossible, uncontrivable, 109,2; 114,19; 116,7; 140,27; 151,23(bis); 162,18
amerês, undivided, 141,18
amphoreus, amphora, 109,8.29; 110,1.2.16; 116,19.22.23.24.26; 117,5.6.9.14(bis)
anairein, eliminate (in argument), 104,11; 105,14; 114,12.13; 123,11; 129,5.13; 130,19; 132,8.34; 133,14(bis); 135,14(bis); 136,9
anaitios, non-causative, 128,20
analambanein, resume, repeat (earlier material), 120,5; 134,24; 145,3; 150,11
analogia, proportioning, 131,6.10

anamenein, await, 146,7
anamphilektôs, unambiguously, 112,6
anankaion, necessary, 126,8; 127,15.19; 128,24; 130,4; 133,18; 137,7.19; 152,15
anankaiôs, necessarily, 140,14; 148,19
anankazesthai, to be compelled, 105,21
anankê (usu. c. *einai*), necessary, 102,2; 105,18; 112,26.28; 118,7; 121,22; 123,26; 124,3; 127,24; 128,29; 132,10.30; 134,24; 135,12.18.25; 136,2; 140,14; 141,5; 146,2; 149,13.17; 151,17; 156,30; 159,21
anaphainesthai, emerge (as a conclusion) 104,2; 106,5; 141,26; 148,11
anaphora, reference back, 143,21
anazôgraphein, picture (mentally), 114,11
anekhein, protrude, 119,10; hold up, 136,22
anesis, release (in physical expansion), 135,12
angeion, vessel, 103,1; 107,7; 108,20; 109,30; 110,16-21; 112,20(bis); 113,18-31.29; 114,4; 115,3; 116,14.15.16; 117,18; 118,21-29; 124,15; 133,27.29; 136,4; see also *amphoreus*, *kaddos*, *kratêr*, *kulix* and *skeuos*
anisotakhês, of unequal speed, 130,20; 132,14; *anisotakhôs*, at unequal speed, 133,11
antilegein, argue against, 123,2; 124,23; 144,27
antimetastasis, mutual replacement, 111,13; 113,14; 118,19
antimethistasthai (usu. c. *allêlois*), engage in mutual replacement, 102,21; 107,3; 113,16; 117,17; 126,20.24
antimetrein, counter-measure (sc. reciprocate a measurement), 153,13-24; 154,1; 163,24; 164,5
antiperiistasthai, be replaced, 129,20
antiperistasis, mutual replacement, 121,24; 135,21

antistrophê, conversion (of premises into conclusion), 102,9
antitupia, resistance, 130,9
anupostatos, non-subsistent, 124,28
apeiros, infinite, 111,3; 116,13.29(bis).30; 117,4; 129,4.5; 130,12; 134,24; 140,13; 141,24; 148,5; *ep'apeiron*, *eis apeiron / -a*, *ad infinitum*, 105,14; 121,14; 130,14; 135,7; 137,12; 141,25; 152,17; *apeirôs*, 142,12
aphairein, (mentally) abstract, 106,13; 114,26; 129,7; *nôi aphairein*, abstract by reasoning, 114,30; (physically) remove, 138,14
aphairesis, (physical) subtraction, 138,15
aphistasthai, be at a distance, 129,17; leave aside (in exposition), 133,15
aphorizein, define separately, mark off, 148,17; 149,4; 154,5
apodeiknunai, demonstrate, 111,8; 112,25; 113,8; 116,2; 118,1; 129,14
apodeiktikôs, in demonstrative form, 132,4
apodeixis, demonstration, 127,25; 134,24
apodekhesthai, accept (a theory), 149,19
apodidonai, supply (answer, definition, explanation), 102,16; 104,12; 105,20; 106,25; 111,15; 117,28; 132,27; 136,25; 146,17
apokekrimenôs, in separation (of the void), 139,30
apokrisis, reply, 132,22
apophainesthai, claim, 106,24
aporein, raise/state a problem, 102,13; 111,16; 120,22; 127,13; 149,27; 161,28; be at a loss (c. gen.), 132,22
aporia, problem, 105,12; 110,23; 121,21; 127,8.11.18; 136,11
aporos, problematical, 125,9; 127,24.26
apostasis, distance, 160,8.10
araiousthai, be attenuated (materially), 135,13; 137,30
arithmein, count, 147,26.28;

148,13-25; 149,1.3; 150,19.23;
151,19.22; 152,2-11.25;
153,4-20; 160,14-27; 161,1-24;
162,23.24
arithmêtikos, capable of counting,
160,29
arithmêtos, countable, 150,25;
152,22.26.30; 160,23-30; 161,1
arithmos, number, 113,7;
124,19.20; 137,28; 138,6.18;
139,1; 146,24; 148,14-30; 149,7;
151,1.2.3.; 152,6-31; 153,1;
154,13-23; 155,1.18.25;
156,2.6.20.21; 160,23.27;
161,4.8.13.20; 162,5-18;
163,5.16; *eidêtikos arithmos*,
eidetic number, 107,14
arkhê, beginning, first principle,
starting-point, 105,9.10.22;
106,29; 111,12; 115,15; 116,12;
124,24; 129,17.24; 147,13;
150,13; 158,15; 161,5; opp.
peras (limit *qua* terminus),
148,8.9; 151,14; 157,19.20;
163,17; opp. *teleutê*, end,
151,19; 158,13; 164,3
asômatos, incorporeal,
105,5.7(bis); 118,20
asumblêtos, incomparable (in a
ratio), 137,15
to asummetron,
incommensurability, 157,6
asunkhutos, unconflated (sc. not
unified), 113,31; 114,5
athroos, mass, *en masse* (of the
separate void), 123,20; 126,4;
127,31; ; 136,2.12; 139,30; (of
air), 129,24
atomos, (plu.) atoms, 133,11;
undivided, 158,16
atopia, absurdity, 104,26; 132,9
atopos, absurd, 'out of place' (puns
asterisked), 104,14*; 107,26*;
110,20; 111,2*; 116,28.31;
117,6.26; 127,16.33; 130,2;
131,25-33; 132,7; 134,5.30;
141,21.26; 149,15; 162,13
autokinêtos, self-moved, 109,1
auxanesthai, increase, 105,17;
116,3.6; 121,25; 124,6; 162,2
auxesthai, increase, 116,3.4;
121,22.23; 127,3.5; 148,21;
161,7
auxêsis, increase, 111,21(bis);
116,9; 120,12; 123,25;
124,5(bis); 127,1-19; 152,11;
159,16; 161,10.26.31; 162,5;
163,15

baros, heaviness, 119,6; 125,1.10;
130,25; weight, 131,1.20.27
barus, heavy, 103,15; 119,1.7;
125,2.3.14; 130,27;
132,26.30.34(bis); 134,8.13;
136,24; 137,30; 139,23.26.28;
140,2.4
barutês, heaviness, 134,17
bia, force, 122,6; 124,14; 128,29
biaios, forced (sc. movement),
128,29; 129,1
biazesthai, force, 129,16.28; 130,1;
131,26; 133,26
boulesthai, wish, mean, intend,
103,29; 107,13; 111,6;
114,11(bis).16.18; 115,19;
118,29; 120,20; 122,29; 123,12;
126,7; 127,10; 128,15; 130,15;
140,1; 147,14; 157,18
bradus, slow, 130,19.25; 132,25;
142,25; 143,11-16; 152,13;
153,1; 159,22; 160,4.6
bradutês, slowness, 161,25
brakhus, short, 152,29.30

deiknunai, demonstrate, show,
103,27; 104,25; 112,29; 113,8;
114,12; 121,21; 123,1.4.9.13;
125,27; 128,8; 133,12.16.30;
139,20; 155,24.25
deisthai, need, require, 121,17;
129,10; 135,6(bis); 155,27
dekhesthai, receive, 103,2.4;
104,24.26.27.28(bis).29;
105,2.9; 106,17.21; 108,4;
110,1; 111,10; 113,23; 118,15;
123,32; 124,1.11.14; 125,32;
126,1; 133,27; 135,1; 139,1.4.6
dektikos, capable of receiving,
103,17; 125,5; 136,8; 140,4
diairein, divide, 111,28;
112,12.14.19; 120,7.9.11;
122,16; 124,2.15; 132,30;
147,8.12.23; 151,5.27; 157,12-28
diairesis, division, 123,19
diairetos, divisible, divided,
141,25.30; 151,3; 152,23
diakheisthai, expand, 135,13.28
diakrinein, separate, 103,14;
124,16.21; 150,20; 161,3

Indexes 133

diakrisis, separation, 138,14 (opp. *sunkrisis*); 161,4
dialambanein, distinguish, 123,14; 148,11.13; 150,26; 161,3
dialeipein, stop, leave off, 139,14; 147,9
dialêpsis, distinction, 161,3
diamenein, persist (through time), 140,25; 141,27.32; 142,4
diametros, diameter, 156,27; 157,6
diapherein, differ, be distinguished/differentiated from, 103,18; 104,30; 105,2; 107,21; 108,4; 109,24.25.27.28; 110,8; 114,14; 115,3; 118,17; 122,27; 128,11; 125,32; 128,10,12,16; 130,22; 134,22; 135,3; 138,15; 143,16; 144,31; 146,12.14.19.28; 149,25.27; 150,6.17; 151,12; 155,5; 157,15; 161,21; 162,17; 163,1
diaphora, difference, differentiation, distinction, 103,6.10; 111,26; 114,24; 128,28; 129,4.7.8.9; 130,28; 132,16; 138,1; 150,26; 151,4; 157,27; 160,3; 161,25
diaporein, work through a problem, 111,3; 121,22; 140,9
diastasis, direction, 103,7 (= 208b14)
diastêma, extension, 104,17; 106,9-15; 112,26; 113,7-28; 114,2-25; 115-18 *passim*; 122,1-29; 123,5.13; 125,5-30; 126,2-11.22.23; 127,1; 128,7-21; 131,2.14.21; 132,14; 134,2.4; 134,4-19.25(bis).26.27(bis); 135,2-6; 136,8-12; 139,19; 140,21; 141,18; 143,13; 145,21(bis). 22.24; 146,9; 151,21; 159,27; 161,30; 162,1.4; *d. aoriston*, undetermined extension, 106,12-13; *d. kekhôrismenon*, separate extension, 126,22; 128,6-7; 136,12; *d. sômatikon*, bodily extension, 121,29-122,1; *oikeion d.*, (a body's) own extension, 115,27.28; 116,16.17.28
diastizein, undergo division, 141,28
diienai, go/pass through, pervade, 124,7; 127,12; 134,4.6; 135,16

diistanai, place at a distance, 103,16; *diistasthai*, be dispersed, 133,10; *diestanai*, be extended, 104,14.23; 114,29
diorizein, divide, demarcate, separate, 103,10.19; 124,20; 140,7.24; 147,6.26.28(bis); 149,1; 151,5; 152,17.30; 159,20; determine (by argument), 108,21; 115,15; 125,5
dogma, doctrine 104,18; 124,23; *ta agrapha dogmata*, unwritten doctrines (Plat.), 106,22.23
doxa, belief, doctrine, 102,6; 113,9; 115,12; 116,10; 121,15; 122,32; 123,3.8; 125,25; 142,17; *koinai doxai*, standard beliefs, 122,32; *palaia doxa*, traditional belief, 102,5-6; 113,9
dunamis, power, capacity, 103,7; 107,5; 132,31; *dunamei*, potentially, 103,17; 121,5; 137,26; 138,3.20.23.24; 140,1; 157,16.26; 160,17.18.24.25.29; *kata dunamin*, potentially, 120,6(bis).8.14.16
dunasthai, be able, can, 103,22; 104,23; 108,21.27; 109,15; 114,10.22; 117,29; 124,27; 126,20; 127,32; 138,20; 143,1.7; 147,16; 151,7; 156,20; 157,8.17; 158,8; 162,21
dunatos, possible, 102,19; 110,15; 114,12.18; 115,34; 124,26; 133,10; 134,29; 137,10; 140,27; 141,13.17; 143,2; 147,12; 160,20.22; 161,4.5
dustheratos, hard to pin down (metaph.), 111,5

ê, introducing the answer to a question, 104,13*; 107,9*; 116,1; 150,3*; 152,1; 154,4; 155,25; 158,10*.11*
êdê, in inferences (thereby, *ipso facto*), 118,21; 124,1; 125,30; 135,14; 137,18; 156,3.4; 163,7
eidos, form (opp. matter), 105,11; 106,4.11.7.17.20.22.27; 107.13-29; 108,1-14; 112,26.29.30; 113,1-5; 118,7; kind, 111,21; 162,15.16; species, 120,26
einai, be, exist, *passim*: with a participle in a periphrasis for

the present tense, 106,16; 119,30; 123,15; 139,3-4; *to einai*, the being (of something), 102,4; 134,11; 153,11; 155,22; 156,31; 157,28; 162,12; *ta onta*, things that exist, are the case, 102,6.8; 104,1.8; 109,19; *to ti ên einai*, essence, what it is to be, 109,25; 113,5
eisagein, introduce (a doctrine), 103,25(bis); 127,9.18; 129,13; 137,16; 139,5
ekhein, have, hold, possess, *passim*; occupy (a place), 122,28; 126,26; (c. infin.) be able, 132,29; *ekhesthai*, be in succession, 122,25; 140,8
eklambanein, understand (cf. LSJ V.1), 125,2
ekpurênizesthai, be squeezed out, 126,29.30
ekthlibesthai, be squeezed out, 124,13; 126,28; 137,3.4
ekrhein, flow out, 102,21; 113,29
enantios, contrary, opposite, 104,10; 137,24; 142,23; 157,8
enantiôsis, pairing of opposite natural qualities, 137,25; 138,4
enargeia, direct observation, 163,25
endekhesthai, be possible, 109,28.31; 111,29; 116,11; 128,25; 136,2; 137,9.23; 152,7; 153,8.9; 156,5.26; 160,30; 161,2
enêremein, be at rest, 118,2.3
energeia, activity, 144,17; 159,9; *energeiâi*, actually, in actuality, 123,13; 125,18.20; 137,26; 138,4.24; 157,16.17; 160,18-25; *kat'energeian*, actually, in actuality, 120,6.8.10.15; 121,5; 125,24.30
energein, perform an activity, 144,14
enestanai, be in present time, 141,2; 158,21
eniautos, year, 140,13.29; 163,18.24
enistasthai, object to, oppose (a theory), 123,8
enkerannusthai, be mixed in, be intermingled, 126,5.30; 136,13; 137,9; 140,1
ennoein, conceive, reason, 111,13; 113,27; 114,12; 137,20; 143,27; 145,8; 148,4
ennoia, concept, notion, 111,6.24.30; 115,15; 144,26; 153,22; 163,2; *koinai ennoiai*, common (sc. widely-shared) notions, 115,14-15; *prôtai ennoiai*, primary notions, 111,30
entelekheia, entelechy, 146,20
epekhein, occupy (a place), 104,16
epekteinesthai, be extended (in volume), 139,17
ephelkesthai (c. dat.), be committed, 128,9
epideiknunai, demonstrate, 103,5; 114,21; 121,15; 123,8; 124,28; 125,15; 126,6; 127,26; 129,2; 132,25; 137,22; 143,12; 152,12
epididonai, develop, 138,28.29
epidosis, development (sc. growth), 127,2
epinoein, conceive, 110,15; 114,25
epinoia, conception, 103,23; 113,13; 125,31; 145,1; 157,18; 162,21.25; 163,5
epipedon, plane, 106,11
epiphaneia, surface, 104,27.28.31.32; 105,1.2.3; 107,10; 109,13-16; 113,5; 114,3.8; 116,9; 117,9; 119,22; 141,31; *koilê epiphaneia*, 'hollow' (sc. inner) surface of a container, 112,22; 113,9.18-19.22.31; 114,19-20; 116,25; 118,24; *mathematikê epiphaneia*, mathematical surface, 128,13
episkeptesthai, investigate, 102,2; 119,29; 127,30; 143,7; 146,13.28; 163,11
episkopein, investigate, 109,18; realise, 128,25
epistêmê, knowledge, 109,17(bis); 111,29
episurrhein, flow forward, 129,23
epiteinesthai, be intensified (of qualities), 139,15
epôthein, supply a forward thrust, 129,22
êremein, be at rest, 118,11; 128,27; 130,4; 151,11; 156,6.10.13.15.24.27; 160,19
êremia, rest, 143,26; 145,7; 155,30.31; 156,1.5-28; 160,19

ergon, task, 144,15; activity, 163,7; result of activity, 161,8; end-product, 149,2
to eskhaton, extremity (of a container; i.e., place), 112,10; 119,3.12; 121,18; 129,6; *ta eskhata*, (physical) extremities, 112,18.21.27; 116,19.24; 118,24; *eskhaton (-a) tou periekhontos*, extremity (-ies) of the container (df. of place), 112,15.24.27; 115,13; 116,20; 117,7
eudaimonia, happiness, 108,18
eudiairetos, easily divided, 131,3
eulogos, reasonable, 150,13; 163,2; *eulogôs*, reasonably, 122,13.14.23; 153,1.23; 155,17; 161,17
eutheia (sc. *grammê*), straight (line), 160,1
exairein, remove, extract, 114,1,7; 136,4; 144,9
exetazein, examine, 122,32; 124,25; 126,17; 127,31; 133,15; 142,17; 163,7
existanai, displace, remove, 155,18; 159,5; (pass.) be displaced, be removed, 108,3; 125,26; 133,18; 137,13; 159,5; (middle) [suddenly] emerge, 158,24
exousia, licence (in linguistic use), 109,4

genesis, coming into being, 138,15.16; 141,13; 157,9; 159,3.7.9.16; 161,10; 162,7; 163,15
genos, genus, 104,13; 108,10(bis); 120,26; 162,15
ginesthai, (45/94) come to be, come into being, come into existence; passim; *ginesthai en*, come to be in (i.e. enter [a place]), 104,16; 107,3.27; 108,3 ; 115,26.27.28*; 116,18.21; 117,4.8.19; 120,13-14; 121,25
ginôskein, get to know, 106,30(bis); 118,12
gnôrimos, knowable, recognizable, 132,18; 163,16; 150,27
gnôrizein, get to know, recognize, 108,25; 148,7; 150,18; 160,2
gnôsis, knowledge, 111,26

grammê, line, 104,28.31; 105,2; 131,4; 140,17; 141,21.23.31; 147,1.13; 148,6; 151,12.20.21.25 (ter); 152,19.20.22; 157,15.16; 160,12; 163,9.10; *eutheia grammê*, straight line, 160,1; *mathêmatikai grammai*, mathematical lines, 157,22
graphein, write, 112,3; 145,15; 147,8-21; 150,10; draw, 103,22

haphê, touch, 125,8.13
haplous, simplistic, 142,18; unidirectional (of time and the celestial revolution), 142,23.24; see *sôma*; *haplôs*, without qualification, unqualifiedly, 118,25; 121,7.9; 124,30; 126,13; 132,19; 148,8; 151,7; 152,18; 156,16.17.29; comp. 141,21
haptesthai, touch, be in contact, 112,15.18; 114,22; 120,9; 121,18; 122,6.16.20; 129,17.20
haptos, tangible, 124,32; 125,6.14.19
harmozein (c. pros), apply to (of an argument), 128,6
hêgoumenon, antecedent (in a conditional), 116,3
hepesthai, follow (as a logical consequence), 125,2; 128,22; 132,8.9; 135,23; 140,4; 150,14; 156,25
hexis (transliterated), 107,5; 111,1; 130,16.17; 155,32; 160,14
histasthai (c. *pros*), address, confront (a problem or doctrine), 124,24; 127,26
to holon/ta hola, the whole cosmos, 119,4.24
horân, see, 115,5; 126,22; 127,33; 130,20; *hora*, see (injunction to hearers), 104,9; 134,24
horisomos, definition, 102,16; 148,25; 149,6.17; 154,22
horizein, bound (of form in relation to matter), 106,6.10.15; 113,6; define, be defined (with reference to temporal parameters), 143,12.14.18.20; 145,5; 146,18.20; 147,3; 148,4; 151,21; 152,25; 154,22; 155,22; 158,4.7.9; 159,16; 163,11.20.21

hormasthai, to start (from the premises in an argument), 126,10
hormê, impulse, 115,11(bis)
horos, boundary, 157,14; 160,8; 163,19
hugiês, valid (argument), 102,6; 113,23
hulê, matter, *passim*, 105,10; 106,6-28; 107,10-24; 108,2.3.14; 112,26.29; 113,6; 118,2-7; 122,16.19; 125,20-23; 137,24; 138,2-30; 139,4-22; 140,3; 159,15; *hulê hupokeimenê*, underlying matter, 106,6; 153,19; 159,15
huparkhein, exist, be the case, 114,12; 123,5.30; 125,15; 126,3.4; 129,12.27; 130,26; 134,10.27; 135,18; 141,24; 146,11; 149,20; 151,10; 161,21; 163,4; belong to, be a property of, 103,14.30; 105,17; 110,11; 111,5; 115,14; 117,22; 143,17; 145,17.26; 146,8.9; 162,19; it is possible, 116,2; 129,27; 133,5; *ta huparkhonta*, properties (syn. *sumbebêkota*), 106,8; 134,16.29
huparxis, existence, 104,15; 142,8; 150,28; 155,27
hupexisthasthai, be displaced, 133,20; 32
huphistasthai, to subsist (sc. exist as a substance), 114,19.27; 140,16(bis). 25; 160,24; c. *kath' auto* (pleonasm), 114,27
hupokeisthai, underlie (be a substrate for), 159,27; be assumed, 111,14; *hupokeisthô*, let it be assumed, 131,20; 147,9; 151,17; *to hupokeimenon*, substrate, 106,16; 120,5; 133,7; 138,16.28; 139,18; 146,13.16.22.23.26; 159,23(bis); 157,21; (plural), 120,15; 145,19; *hupokeimenôi*, in substrate (opp. *logôi*, in definition), 122,26-7; 146,13-14; 150,4.24-5; 157,25; *kata to hupokeimenon* (opp. *logôi*), 150,6-7
hupolambanein, believe, 102,6.20; 103,8; 104,24; 112,30; 114,22; 115,9; 122,26; 123,4; 124,31; 126,6; 127,29; 137,7; 141,2; 144,19.25.31; 145,6
hupolêpsis, assumption, 102,6
hupomenein, await, 127,21; stay behind, 114,2; persist, 145,24(bis); 146,8; 151,14; 157,15
huponoein, infer, 115,4; interpret, 141,3
hupostasis, substantiality (opp. in conception), 163,1
hupothesis, assumption, hypothesis, 110,19; 114,9; 125,13; 132,8; 133,1
hupotithesthai, assume, hypothesise, 107,15; 110,17; 114,7.14; 142,29; 143,3
husteron, later; *to husteron*, the after, *passim*; see *proteron*

idiôma, uniqueness, 152,10
idios, special, unique, 105,24; 113,19; 115,30; 120,30; 128,2; 130,29; 135,4; see under *kinêsis* and *phora*
isotakhôs, at an equal speed, 132,28.34; 133,12
isousthai, be made equal (in number), 153,21

kaddos, pot (sim. amphora), 112,22; 113,9; 118,26
katakermatizein, split up, 136,10 (see n. 239)
katalambanein, overtake (by acceleration), 133,13
katametrein, measure out (in units), 140,20; 154,5-11
kataskeuazein, establish (a thesis), 126,16; 133,14; 142,4
katastêma, (psychological) state, 145,5
katêgoreisthai, be predicated, 158,9
katekhein, occupy (a place), 105,18; 116,16-23; 117,11; 134,13
kath' hauto/kath' hauta (and variants), *per se*, 102,4.15; 103,20.21; 106,12; 110,5; 111,5.24; 112,1(bis).5.6.7; 113,24; 114,27(bis); 115,1; 116,9.12.31.32; 117,3.13.18.25.30; 118,30; 123,20; 124,11; 128,13.17.22;

133,15; 137,1.5.26; 142,18.20; 152,1.5; 159,3.4; 163,1
katholou, universal (of a proposition), 102,9; *to katholou* (adv.), universally, 143,7
keisthai, c. *en*, to depend on, 108,17.18; be assumed, 112,9; *keimena*, assumptions, 131,11; 137,22
kenos (107/189) (adj.) empty, 105,15; 118,22; empty of meaning, 114,20; 133,16; *(to) kenon*, (the) void, *passim* chs 6-9: e.g. 103,25(bis); 105,14; 113,10; 114,20.23; 115,18; 118,22; 121,12; 122,25-8; 138,13; 139,4-30; *kenon apeiron*, infinite void, 129,4; 130,12; *k. athroon*, mass (sc. separate) void, 123,20; 126,4; 127,31; 136,2.12; 139,30; *k. kekhôrismenon*, separate void, 123,20; 127,32; *k. <paresparmenon>*, disseminated (sc. interstitial) void, 123,16; 135,12; *ta kena*, void spaces (sc. interstitial void), 124,12; 127,14.22; 136,7; 137,3(bis).12
kerameios/kerameous, made of clay, earthenware, 109,8; 110,2; 112,22; 117,17.26; 118,26; 148,27
khaos, 'khaos' (Hesiodic primordial space), 103,29; 104,3
khôra, space, 103,29; 105,4; 106,19.20; 113,30; 126,9.13; 134,20; 135,16.28; 136,6
khôrein, go: *sôma dia sômatos khôrein*, body going through body (sc. total bodily interpenetration), 104,15; 124,8; 126,15.25; 127,1.15-16; 133,7
khôris, separately, 106,30; 107,16; 109,10; 110,19; 115,16; 118,3; 120,32; 125,29; 135,2
khôristos, separable, separate, 107,6.9; 111,12; 120,4; 121,28; 123,13; 125,23.24; 128,15; 136,13; 137,27
khôrizein, separate, 109,26; 113,17.22; 114,8.10.11.28; 123,20; 125,27.30; 126,2.22;
127,31.32; 128,7; 134,18; 135,8; 136,10.12
khronos (194/322), time, *passim*; (i) time (in general; with and without the article), e.g. 140,8-23; 141,16-27; 142,6-31; 143,2-23; (ii) (specific) time-period (references confined to chs 8 and 9), 131,6 (plu.); 131,7.8.9(bis). 10.11.14.15.16 (ter). 17.18(bis).23; 132,10.11(bis).12(plu.).24; 136,31; *kh. mellomenos*, future time, 157,13; *kh. haplous*, unidirectional time, 142,23-4; *kh. hôrismenos*, delimited time(-period), 158,4; *kh. lambanomenos*, identified (sc. delimited) time(-period), 158,9.10; *kh. mellôn*, future time, 142,16; 157,12.13; *kh. parelthôn*, past time, 141,16; 157,2(bis).31; 158,17; *kh. peperasmenos*, finite time-period, 132,11(plu.); 141,28; *khronos tis*, a specific time(-period), 163,17; see also *to mellon*, and *to parelthon*; *en khronôi*, (i) in time (i.e. determinable in principle by time), 104,4; 141,6.17; 142,10; 143,4.13.19; 154,11-24; 155,10-29; 159,14.21; 160,7.9.12(bis); 161,30; (ii) in a (specific) time-period (references from ch. 8 only), 131,19.22.27.29.31; 132,9; 137,14-15
kinein, cause movement, 109,3; provoke (sensation), 118,22; *to kinoun*, the cause of movement ('the efficient cause'), 105,11
kineisthai, change, move, be moved, *passim*: translated 'move'/'be moved' at: 103,15; 107,16.20.26; 109,2; 110,9; 114,3; 115,11; 116,17-30; 118,28(bis); 119,17.18(bis).20.25; 123,25.26.27; 124,8; 128,3.25.28; 129,19.26; 130,1.2.7.8.9.10.23.27.28.30; 131,3.4.6.7.8.14.18.19.21.22.27. 28(bis).29.30(bis).32;

132,4.13.27.28.29; 133,1.2; 143,13 159,25.26.27; 160,2; move (of the human body), 115,10; *kata phusin kineisthai*, engage in natural movement, 107,16-17; 128,29.30; 129,1; 130,1; *para phusin kineisthai*, engage in contra-natural movement, 129,3; *kata topon kineisthai*, change in respect of place, 126,21; 135,15; *kuklôi kineisthai*, move in a circle, 119,15.20.25; *sôma dia sômatos kineisthai*, body moving through body, 123,27-8
kinêsis, change, movement, *passim*: translated 'movement' at: 103,14; 107,22.24; 115,10; 119,20; 124,3.9; 127,32.33; 128,1.6.19.24; 129,12.13.26; 130,5.18.19.20; 132,1.9.10.14.17; 133,2.14.15; 142,24; 159,26; 160,1; (human bodily) movement, 115,10; *k. biaia/biâi*, forced movement, 128,29(bis); 129,1; *k. endotheisa*, imparted movement (sc. by air); *k. homalê*, smooth (sc. uniform) movement, 159,26; 160,1; *k. idia*, special movement (of heavenly spheres), 120,30; *k. kata phusin*, natural movement, 107,17.24; 119,6; 128,29.30; 129,1.2.6.9.15; 137,14; *k. para phusin*, contra-natural movement, 129,1.10.16; *k. kata topon*, change in respect of place, 102,13; 111,18-19; 123,24.25; 126,20; 136,15
kinêtos, changeable, 120,12.17; 122,4; 146,20; 160,15,18; see *phora*
klepsudra, clepsydra ('water-thief'), 123,10; 133,23
koilos, hollow; see *epiphaneia*; *ta koila*, the hollows (i.e. extension within a vessel), 117,5
kosmos, cosmos, 112,7.8.12; 114,28; 119,3; 130,13;
kratêr, kratêr (mixing bowl), 114,23
kuklophoria, 161,6; 163,16

kuklos, (temporal) cycle, 164,2.4; geometrical circle, 139,9.10.12; *kuklôi*, in a circle, see *kineisthai*, *phora* and *pheresthai*
kulix, kylix, 127,22
kuriôs, in a strict sense (sc. primary use of a term); 106,1; 110,25; 111,25; 119,4; 121,9; 127,6; 141,3; 155,2; 157,10.11.29; comp., 139,11.13; *kuriôtata*, in the fundamental (sense of a term), 108,19

lambanein, take on/acquire, 113,30; 129,25; 164,3; *aisthêsin lambanein*, perceive, 144,7; 147,7; understand, 150,3; take (i.e. select, or identify), 131,26; 136,30; 147,12; 149,26; 153,6.23; 154,6; 157,17.18.19.23.26; 158,8.9; 161,15; 163,13; (pass.), be captured, 158,5
legein (87/209), claim, say, tell, *passim*; *kuriôs legesthai*, to be used/spoken of in a strict sense, 110,25; 157,10
leptos, tenuous, 126,29; 130,29; 131,5.9.21
leptunesthai, be attenuated, 138,22
logos, argument, 102,7; 104,10.20; 109,27; 107,11; 116,8; 117,23; 121,14; 124,4; 125,26; 126,17; 127,7.21; 128,6; 129,16; 130,6.20; 134,18; 140,9; 143,5; 149,18; definition, 102,14; 106,28; 108,14; 109,16; 110,4; 111,15; 113,5; 133,31; 145,16; 146,19.21; 150,7; discussion, treatment (of a topic; usu. c. *peri*), 102,5; 106,28; 108,7; 109,25; 122,25.30; 130,28; ratio, 132,29; 136,31; 131,17; 132,11; *kata logon*, in accordance with a formal definition, 137,28; 150,26; opp. *kata to hupokeimenon*, 157,22; *logôi*, in definition, 114,28; 125,24; 135,3; 157,20.23; (opp. *hupokeimenôi*, 'in substrate'), 122,27; 146,14; 150,4.7.25; 157,25; *tôi logôi*, verbally, 162,24; *logon ekhein*: (i) have a

rationale, make sense, 105,12; 162,22; (ii) be in a ratio, 131,12.16.25.33; 132,29; 136,31; *logou kharin*, for argument's sake, 131,29
luein, solve (a problem), 111,16; 121,21; 127,28

mallon de, used to introduce a more precise version of a claim ('rather', 'specifically' etc.), 105,25; 106,18; 109,28; 122,1; 124,20; 132,33; 147,16; 155,26; 161,5
manos, rare, 131,1; 135,9; 136,8.17.19.28; 139,21-30
manôsis, rarefaction, 119,25; 135,11.14
manotês, rarity, 140,3
manousthai, be rarefied, 136,3; 137,23; 139,3
mathêmatikos, mathematical, see *epiphaneia, grammê, theôrêma*
megethos, magnitude, 105,8; 106,6; 116,5; 127,2; 131,28; 134,16; 138,21; 139,7.16; 145,23-8; 146,2-18; 149,10.13; 150,12.14; 152,23.25.26.28; 154,1; 161,29
mêkhanasthai, fabricate (an implausible theory), 138,16
to mellon, the future, 140,23.24; 141,12.14; 142,16; 147,25; 160,8; see *khronos*
menein, remain, remain stationary, 103,1; 106,18; 108,4; 113,16.19.21.30; 114,8.20; 115,18.20; 116,13; 117,13.29; 118,14; 119,14; 120,19; 122,12.14.19; 124,11; 128,5.8; 130,17.24; 132,16; 134,1; 137,5; 138,2.22.29; 145,5; 150,20
meros, part, 104,33; 107,5; 108,8; 109,1.5.7.11.24; 110,2-26; 112,5.14.16; 115,3; 116,20.22.27.29; 117,2.20.23; 120,27.28; 122,14; 128,22; 130,14; 140,15.16.20.21.23.27; 141,31; 142,15.22; 147,3.4.6.13.24; 151,25; 154,5.14; 158,16; *ana meros*, successively, 113,23; *para meros*, in turn, 138,3
metaballein, transform/be transformed, 107,29; 122,8.17; 133,26; 136,4.5; 137,31; 138,3,26(bis); 139,7; 141,10; 144,12(bis); 145,8; 159,6.23; 160,4
metabolê, transformation, 108,1(plu.); 135,25; 137,17; 143,8.23; 145,4.6.7; 159,2.4.14.20.22; 162,7
metalambanein, substitute (a term in a definition), 148,25
metapherein, transfer (analogically), 113,27
metapheresthai, be transported (sc. change position), 116,14.17; 118,26; 135,4; 150,16
metaphorêtos, transportable, 107,7; 112,20; 118,24.25.27.29
metaphorikôs, metaphorically, 106,26
metatithesthai, change position, 117,6
methistasthai, change position, 117,14.15; 119,20; 135,22; be repositioned, 134,5; reposition (intr.), 111,22
metrein, measure, 153,12.24; 154,1; 155,21; 156,7.8.14.24; 162,9; 163,7.9.11.22; 164,9
metron, measure, 113,6; 148,26; 153,15.16.18; 155,30.31; 156,9.19.25.28; 159,15; 163,4-25; 164,4.6; dimension, 115,34
monadikos, monadic, 152,9; 153,19; 154,18; 161,20
monas, unit, 124,21; 148,30; 151,1.3; 152,18.19; 153,21; 154,18; 163,9(bis)
monê, stability, 128,5
morion, part, 108,27(bis); 109,2; 116,17.23.25.31.32; 122,15; 119,21; 126,24.26; 131,30; 138,8; 140,19; 151,24; 158,20
morphê, structure (syn. form), 106,17; 107,24; 113,1; 118,2; 137,29

noein, conceive, think of, 103,22.23.26; 106,8; 113,8; 114,27; 115,1; 126,1; 134,22; 144,24.27; 145,1; 146,28; 147,2.5; 148,2; 149,24; 106,8; formulate (a belief by reasoning), 102,6

noêsis, conception, 123,4; 157,24
noêtos, intelligible, 105,7
nous, intellect, 160,22; 161,7; mind, 148,2; *tôi nôi*, by reasoning, 114,30
to nun, the now, 140,20-26; 141,3-32; 142,4; 144,8.10; 147,15-29; 148,3.4.7; 149,23.26.27; 150,6-30; 151,1-27; 152,3.25; 154,17; 157,10.15-30; 158,3-21; 159,18; 160,9.10; 161,12.13; 162,15.20; *to akribôs nun*, the now in a precise sense, 141,3; 157,32; *kuriôs*, the now in a strict sense, 157,10-11,29; *en platei*, the now in an extended sense, 157,10; *to paron nun*, the present now, 147,15.17; 158,3.4.16.20; *to parelthon nun*, the past now, 158,20

oiesthai, think, believe, 102,8; 103,30; 110,23; 111,19; 115,21; 123,6.21; 124,4; 126,16.18; 132,25; 135,9; 144,5.24; 149,4.15; 160,27; 161,20
oikeios, own, proper, inherent, see *diastêma*, *phusis*, *topos*
oikhesthai, to be eliminated (of a thesis), 129,13
onkos, volume, 115,36; 124,10; 127,3.5; 133,18-32; 134,15-33; 135,10.11.26; 136,3.6; 137,10.17.20.22; 138,9.22; 139,9
onoma, name, 106,26; 149,17; 163,19.26
onomazein, name, 107,16; 109,5; 162,17
ôthein, push (forward), 129,20.23.26.28; cogn. *epôthein*
ouketi, in emphatic negations (usu. adversatively), 109,13; 114,23; 115,35; 117,7; 119,15; 121,20; 125,11; 130,11; 133,4; 134,9; 149,22; 152,31; 153,2; no longer, 127,6; 151,9
ouranos, the heavens, 111,19; 119,3.17; 120,4.15.21; 142,11-24; 163,28; the cosmos, 105,26; 117,22; 121,15; 123,15.20; (plu.) universes, 142,31
ousia, a substance, 102,4; 115,25 (opp. *sumbebêkos*); 132,31; 137,28; 142,16; 150,6.28; being, essence, 142,7; 143,9; 149,2; 153,10; 155,22; 156,31; 159,6

pakhunesthai, be compacted, 153,27
pakhus (opp. *leptos*), compact, 126,5.31; 130,29; 131,2.23
to pan, the All (sc. the whole cosmos) (see, *ta hola*, *to holon*, *ouranos*); the whole heavens (*ho pas ouranos*), 142,22.27.30; 163,16
paradekhesthai, accept, condone (a doctrine), 115,19; 116,2; 127,20.25; 135,25; 163,8
paraspieresthai, be disseminated (of interstitial void), 123,16; 135,12; 136,8
to parelthon, the past, 140,24; 141,12.15; 142,15; 147,25; 160,8; see *khronos*
paremplokê, interlacing (of atoms), 123,18
paristanai, present (problems), 119,26; (pass. c. dat.), engage with (a topic), 104,19
paskhein, be affected, 110,12; 122,10.12; 145,9; 155,10.20
pathos, affection, 107,5; 110,14.17.19; 111,1; 134,8.17.20; 135,3; 140,6; 145,17; 154,14; 159,24; 160,14; 162,17; *pathê sômatika*, corporeal affections, 161,6
pêgnusthai, be compacted, 139,3
perainesthai (usu. *peperanthai*), be limited, be finite, 121,13; 131,17; 132,11; 141,28.31; 158,6
peras, limit (of a body, or of a temporal sequence), 106,1-5; 107,10; 110,25.26; 112,17.24; 113,2-6.20; 116,9; 118,8.21-30; 119,1.2.15.30; 121,23.25; 122,1.3.21; 128,9.12.13; 140,15; 141,28.31; 147,13; 151,20.26; 151,1.4(bis).5.6; terminus (of a temporal process; usu. opp. *arkhê*, beginning), 148,8.9; 151,13; 157,19.20; 163,17; outer limit (of a race-course), 146,1; terminus (of a movement), 160,3; *peras tou periekhontos*, limit of the container (df. of place), 110,26;

112,24; 113,8 (*perata*); 116,8; 117,1; 118,8; 119,1-2; 128,9; see *eskhaton*
peratoun, limit, delimit, 106,2.5
periekhein, contain (of place)/be contained (of something in place), 104,33; 105,3.27; 106,1; 107,10.11; 110,26; 111,7.8; 112,2-27; 113,1.3.4.8; 114,3; 116,8.20; 117,1.7; 118,8-29; 119,2-29; 120,10.16.30; 121,1.3.15.20.24; 122,6.10; 123,15; 128,12; 136,6; 141,1.2.4; 154,24; 155,1.28; 157,1; be included (in a definition), 108,14; surround (of extra-cosmic void), 123,15.20
to periekhon, the container (used of place), see *peras*
perigraphesthai, be circumscribed (by limits), 113,19; 143,9
periodos, circuit (of time), 164,3
peripheresthai, be carried round (in a celestial revolution), 120,18
periphora, (celestial) revolution, 118,11; 142,19.21.22(bis).30; 163,24-29
peripiptein, confront (mentally), 125,16; 136,11; 148,8
peritteuein, extend beyond, 104,33; 155,22.24.27
phainesthai, appear, 106,8; 119,5; 145,5; be manifest, 124,10; 127,9; 133,16; 162,12; emerge (as the conclusion of reasoning), 104,23; 163,14 (see *anaphainesthai*); *enargôs phainomena*, self-evident phenomena, 133,30
phanai, say, 106,19.21; 107,15; 108,4; 109,3.8; 112,14; 114,25; 116,12; 120,20; 125,7; 135,38; 137,1; 144,27; 149,6.11; 160,26; 161,27
phantazesthai, be imagined, 106,10; 118,22; imagine, 145,10
phenakizein, dupe, 112,30; 113,30
pheresthai, be carried (of a body moving to its proper place), 103,7; 107,17.18; 119,7; 122,5.8.9.13.21.22.24; 129,26; 134,14; be carried (in the void), 128,5; 130,5.13.14; 133,11; 136,22; (of the void itself), 136,29; be carried (of an inanimate object being moved), 130,23.26.30; 132,16.21.24.30; 133,6; 136,21.25; be in motion (as determinable by time), 150,15-27; 151,6-24; *kata phusin pheresthai*, be carried naturally, 129,26; *kuklôi pheresthai* (= *peripheresthai*), be carried round in a circle, 119,22
phora, motion, locomotion, 111,21; 118,11; 119,4; 123,25.26; 129,26; 136,23; 140,5; 150,18.27; 151,6.25; 156,4.9; 161,13.25; 163,15(bis); 164,4; *ph. idia*, unique motion (of elements), 128,2; *ph. kata topon* (pleonasm), motion in respect of place (sc. locomotion), 159,25; 161,10; *kata ph. kineisthai*, to change in respect of motion, 118,10; *kata ph. kinêtos*, changeable in respect of motion, 120,12.17; 122,5; *kuklôi ph.*, circular motion, 119,12
phôrasthai, be detected (by observation), 133,22
phtheirein/phtheiresthai, destroy/be destroyed, cease to be, 104,6.7; 108,1; 115,29.30; 140,26; 141,5-22; 155,13.14; 158,26; 159,10
phthisis, decrease, 111,21; 123,25; 161,10
phthora, destruction, ceasing to be, 108,2; 138,16; 155,17; 157,9; 159,3.11.13.15; 162,7
phusikos, natural, 137,25; pertaining to natural philosophy (c. *hupomnêma*), 104,21; 119,26; *ho phusikos*, natural philosopher, 102,2.11; 122,25; see *sôma*
phusis, Nature (personified), 103,17; a natural reality (sc. natural substance), 102,12; 109,26; 138,1.11; nature (of something), 109,17.19; 111,5; 112,6; 123,2; 124,19; 128,1; 139,5; 148,11; 150,1; 159,14; (*têi*) *phusei*, by nature, naturally, 103,14; 104,4-5;

126,6; 128,2; *ph. oikeia*,
(something's) own nature,
128,1; 163,3; *ph. sômatikê*,
corporeal nature, 125,18; *hupo tês phuseôs*, through the
agency of nature, 159,16;
163,6; *kata/para phusin*: see
kineisthai, kinêsis
pilêsis, compression, 121,24;
126,31; 133,28; 137,5.7.9;
137,9; 139,20
pilousthai, be compressed, 124,10;
126,29; 133,26; 135,13;
137,1.6.12.29
plêrês, full (sc. solid), 116,14;
123,7.27.30; 125,7.12; 130,8;
131,12.31.33; 132,1.12.13.30;
137,1
poiein, do, make, produce, 107,13;
115,23; 116,1.15; 117,23; 119,8;
122,10; 123,9; 124,14;
125,21.23; 127,25.31;
128,6.13.22; 134,32;
136,10.12.17; 138,15.16;
142,7.8.19; 144,9; 147,4.24;
149,1.2.9; 151,2.3.5; 152,16;
159,13; 161,4.24; 163,21;
represent (in literature),
103,28; 142,28: *arkhên poieisthai* (= *arkhesthai*),
begin, 105,22; *marturion poieisthai* (= *martureisthai*),
provide evidence, 124,14
poiotês, quality, 114,30; 115,3.6.7;
118,7; 125,10; 138,5
pragma, (physical) object (content
of a place), 107,2.12.25; 111,7;
119,1; 125,23; subject (of
discussion or analysis), 123,2;
130,18
proêgoumenôs, directly (opp. *kata sumbebêkos*), 153,12; 162,9.10
proistasthai (c. gen.), be a
proponent of, 123,3
to prokeimenon, the proposed
(object of analysis), 124,25
prolambanein, anticipate, 128,23;
ta prolabonta, presuppositions,
102,17
proodos, progression (to infinity),
121,14
proôthein, push forward, 135,18
prosaptesthai, be attached to (of
matter in relation to place),
106,9

proseinai (c. dat.), belong to (as a
property), 102,15 (syn.
huparkhein, 208a34)
prosekhesthai, to align oneself
with (a philosopher), 149,4
prosekhês, proximate (of place),
106,4
prosekhôs, proximately, 106,3
prostithenai/prostithesthai,
add/be added, 104,10;
116,26.30; 124,18; 125,4.6;
127,5; 128,21; 133,24
proteron, earlier; *to proteron*, the
before, usu. c. *to husteron*, the
before and the after
(temporally and spatially),
passim chs 11-14, e.g. 145,27;
146,2.4-5.13.15.19; 152,26;
153,2.8.10-11; 156,20-1;
160,6-7.9.11;
161,1-2.9.14.26.29.30;
162,3-5.8.13-14
prôtos, primary (sc. fundamental),
102,14; 106,4; see *topos*; *prôtôs*,
in a primary sense, 106,5;
108,25; 109,14(bis).17.19.20.26
proüparkhein, pre-exist, 139,2;
155,18
psukhê, mind, soul, 108,25;
109,3.17.18.20; 120,13.21;
145,5; 147,29; 148,7 (= *nous*,
148,2); 160,21.22.26; 161,4-8;
163,2.4
puknos, compact, 135,9; 137,30;
139,21.23.25.26
puknôsis, compacting, 119,25;
121,24; 135,10.14; 137,16
puknotês, compactness, 140,4
puknousthai, be compacted,
124,13; 126,27.28; 137,3

rhêgnusthai, be broken, 133,9;
136,4
rhepein, be on a trajectory, 133,9;
136,26
rhipsis, throwing, 129,19
rhiptein/rhiptesthai, be thrown,
129,18.20.21
rhopê, trajectory, 130,21
rhumê, forward rush, 129,21.22

sêmainein, signify, 103,29; 123,12;
124,25.26.27.29; 125,12;
149,5.14.17; *kuriôs*

sêmainesthai, be signified in a strict sense, 155,2
sêmeion, (evidentiary) sign, 148,18; point, 104,29; 105,3(bis); 141,23; 145,28; 146,4.6; 147,12.18; 151,19; 157,14.15; 158,7
skemma, investigation, 102,14; 106,21
skeuos, vessel (syn. *angeion*), 109,8.9.10; 110,3.4; 113,20; 117,29
skhesis (c. *pros*), relation, 103,8.21
sôma (83/273), body, *passim*; *sôma aisthêton*, perceptible body, 123,6; *sômata hapla*, uncompounded bodies (sc. elements), 128,1.15; *sôma kinêton kata phoran*, body changeable in respect of motion, 120,12; 122,4; *sôma phusikon/sômata phusika*, natural body/bodies, 102,10.13; 103,14; 128,8.14.15; 129,6; *sôma dia sômatos khorein*: see *khôrein*
sôzein, preserve (phenomena), 115,31; 127,10.20; 129,2.3.13.18.
sphaira, sphere, 106,13; 119,21.22; 120,30; 121,2.7; 130,27; 131,3.5.6; 132,28; 142,10.14.16
stadiaios, a stade in length, 131,4.21.28; 132,2.3
stadion, stade, 131,21.24; 132,1; 146,1; 164,6
stasis, stationary condition, 143,26
stereisthai, be deprived, 103,26
sterêsis, deprivation, 129,9; 139,22; 155,31.32
stigmê, (geometrical) point, 104,30; 121,28(bis); 147,7; 150,12-25; 151,12.13.25; 157,23.27
stoikheion, element, 105,5(bis).6; 118,23; 119,10; 138,17
sumbainein, happen, turn out, result, be a consequence (c. infin.), 115,2; 128,24; 132,21; 136,15; 141,8; 144,6; 153,19.24; 159,5-6; (absol.) 133,2; 135,20; 149,15; 153,19; 155,5.10; *sumbebêke*, be incidental to, 110,10.13; 152,4; 153,10; 156,3.4; 160,30; *to sumbainon*, the consequence, 129,30

sumbebêkos, incidental [property], 107,8; 115,24; 145,17; 156,3; 161,16; *ta sumbebêkota*, incidental properties, 106,13; 115,35; 116,1; 134,11.19.21.30; 154,19; *kata sumbebêkos*, incidentally, 107,2.9; 110,6-12; 111,24.26; 117,1.20.29; 118,25; 120,13-32; 121,6.8; 153,12; 156,8.10.29; 159,7
sumballein (intrans.), coincide, 112,23
summetros, commensurable, 156,27; *to summetron*, commensurability, 157,6
sumparekteinesthai, be coextended, 141,17
sumperipheresthai, complete a (celestial) revolution with, 142,27
sumphuesthai, be naturally fused together (of parts in wholes), 122,19
sunanairein, eliminate jointly (by argument), 104,5(bis)
sunartasthai, be co-dependent (sc. in a relation of mutual implication), 144,22.26; 145,1
sunduazesthai, be coupled (of matter with form), 137,28
sunekhein, hold together (sc. render continuous),151,5.12; 157,12.24.28; *hê sunekhousa hexis* (Stoic), the cohesive *hexis*, 130,16
sunekhês, continuous, 112,13.15.18; 120,7.9.10; 121,4; 123,15; 140,19; 144,1; 145,21-27; 146,8; 147,2; 149,1,3; 151,4.6.9; 152,14.24.30; 158,2; 161,9; *sunekhôs*, continuously, 147,9
suneteroiousthai, be altered along with, 155,26
sunêtheia, ordinary usage (of language), 109,4
sungenês, kindred (of proximate body), 122,6.8.9.13.14.22
sunistasthai, be compounded (from), 105,5.11; be involved with (a theory), 123,23
sunkheisthai, be conflated (sc. unified), 114,3; 136,13
sunkineisthai, move together with, 107,25

144 Indexes

sunkrisis, confluence (opp. *diakrisis*), 138,14
sunthlibesthai, be squeezed, 136,6; 137,8
sunuphistasthai, subsist along with, 144,22; 146,10
sustellein, contract, 135,13.15; 137,3.8; 138,23
sustolê, contraction, 135,11; 139,20
suzeugnusthai, be yoked together (sc. in a relation of mutual implication), 144,22; 160,16

teleutê, end (of a temporal sequence; opp. *arkhê*; cf. *peras*), 151,19; 158,13; 164,3
telos, end (sc. purpose), 108,17
temnein, cut into segments, 147,23; 152,25
thaumastos, odd, surprising, 104,1; 117,23; 134,1; 153,18
thaumazein, be surprised at, think odd, 108,11; 148,24
theôrein, inquire into, investigate, 102,15; 111,15; 122,30; 155,17
theôrêmata mathêmatika, objects of mathematical study, 103,19
theôrêtikos, capable of inquiring, 102,12
theôria, inquiry, 102,5; 118,13; 162,25
thesis, position, 103,16; 132,8; 146,1.15; 147,3.4.8; 149,10.12; 151,14; 162,4.10; posit (sc. conclusion of an argument), 102,9; positing (of a concept), 123,16

tithenai, posit, 124,22; usu. *tithesthai*, 113,10; 117,1; 122,31; 123,5; 124,29; 125,31; 126,9; 128,27; 132,4.15; 145,18; (pass.), be positioned, 128,4; c. *en*, be inserted, 133,17.19.32
tode ti, this something (sc. substance), 125,17*; 150,27-8; 151,7
topos (169/273), place, *passim*; *oikeios topos*, (a body's) proper (sc. natural) place, 107,17.20; 122,9.13.21; 129,27; *ho prôtos topos*, primary place, 105,22-4.26; 111,9.11; 112,12.21 (plu.); 118,15; *ho kuriôs topos*, place in a strict sense, 106,1; 110,25
tropos, sense (of a term), way (of proceeding), 108,7.18.22; 109,21; 110.11.22; 121,9; 122,17.30; 123,16; 124,10; 125,17; 140,7; 143,18.22; 144,24; 146,10; 153,19; 154,14.23
trophê, nutriment, 116,7; 124,6; 127,6-17

zêtein, inquire into, investigate, 107,23; 108,21; 114,9.10; 125,24; 140,24; 143,26; 146,21.27; 162,14
zêtêsis, inquiry, 124,25
zôion, animal, 108,10.11.13; 109,1.4.7; 110,8; 112,7.8; 156,10; 161,7

Index of Passages Cited

Omitted from this index are references and cross-references to Aristotle *Physics* 4, and Themistius' paraphrase of that text. For details of editions and abbreviations, see the Bibliography above.

AESCHYLUS
F139 (Nauck; Radt): 91 n. 260

AETIUS (Diels, *Doxographi Graeci*)
1.18.1 (*Dox. Gr.* 315): 82 n. 122;
  1.19.1 (*Dox. Gr.* 317): 78 n. 53;
  1.21.1 (*Dox. Gr.* 318): 99 n. 380

ALEXANDER OF APHRODISIAS
de Anima **4,23-4**: 78 n. 48
de Mixtione **chs. 5-9**: 77 n. 28; **ch. 16**: 89 n. 223
Quaestiones **2.12**: 89 n. 223; 97 n. 346
ap. Simplic. *in Physica* **286,6-10**: 93 n. 296; **526,16-18**: 76 n. 11; **530,19-24**: 89 n. 217; **589,5-8**: 85 n. 164; **679,12-37** (Usener, *Epicurea* no. 279): 94 n. 318; **759,18-760,3**: 109 n. 533

ARISTOPHANES
Clouds **1**: 101 n. 403

ARISTOTLE
Categories **1a1-6**: 99 n. 381; **2a1-2**: 86 n. 180
de Anima **3.10, 433b14-19**: 83 n. 140
de Caelo **1.7**: 87 n. 192; **1.9, 278b21-4**: 88 n. 210; **4.3**: 87 n. 200; **4.5, 312a12-13**: 78 n. 59
de Generatione Animalium **3.11, 762a13-14**: 108 n. 518
de Generatione et Corruptione **1.3**: 87 n. 200; **1.5**: 91 n. 269; **1.5, 320a1-5**: 79 n. 68
de Partibus Animalium **1.2, 642b5**: 104 n. 456
Metaphysica **2.1, 993b5**: 88 n. 207; **5.7, 1017a26**: 86 n. 180; **12.8, 1074a35-8**: 110 n. 537

Meteorologica **2.8, 367a11**: 96 n. 339
Physica **3.1, 201a10-11**: vii; **3.4, 203b22-30**: 87 n. 190; **3.5**: 87 n. 192; **3.5, 206a2-3**: 75 n. 4; **3.7, 207a35-b1**: 78 n. 59; **5.1, 225a34-b3**: 5 n. 14; **5.6, 229b29**: 109 n. 525; **6.3, 233b33-5**: 106 n. 490; **8.1, 251b17-28**: 100 n. 390; **8.8**: 106 n. 483; **8.10, 266b27-267a12**: 92 n. 285
Fragments (ed. Gigon) **F93** (Simplic. *in Phys.* 453,25-30): 79 n. 63; **F96,2** (Themist. *in Phys.* 106,4-26): 78 n. 50; **T22.10** (Themist. *in Phys.* 140,8-142,10): 98 n. 360

PSEUDO-ARISTOTLE
Problemata **25.8, 938b14-24**: 89 n. 225

CLEOMEDES
Caelestia (ed. Todd [4]) **1.1.20-149**: 88 n. 214; **1.1.33-8**: 94 n. 320; **1.1.36-8**: 94 n. 320; **1.1.91-5**: 93 n. 296; **1.1.98-9**: 93 n. 296; **1.1.132-8**: 87 n. 191; **1.5.105**: 90 n. 246; **1.6.103**: 90 n. 246; **1.7.50**: 84 n. 150; **2.1.275**: 90 n. 246; **2.1.401-3**: 83 n. 133; **2.2.6**: 90 n. 246; **2.5.92-101**: 99 n. 370; **2.6.114**: 99 n. 378

EPICUREA (ed. Usener)
No. 274 (Themist. *in Phys.* 123,15-22): 88 n. 214; **No. 279** (Simplic. *in Phys.* 679,12-37): 94 n. 318

EUDEMUS (ed. Wehrli)
F78 (Simplic. *in Phys.* 754,6-16): 77

n. 37; **F80** (Simplic. *in Phys.* 595,3-15): 86 n. 178; **F88** (Simplic. *in Phys.* 732,26-733,1): 105 n. 468; **F90** (Simplic. *in Phys.* 754,6-16), 108 n. 514

HESIOD
Theogony **116**: 76 n. 14

HOMER
Iliad **1.500-1**: 80 n. 73; **6.319**: 103 n. 436; **8.186**: 103 n. 436; **8.494**: 103 n. 446; **17.514**: 79 n. 73; **20.435**: 79 n. 73
Odyssey **1.267**: 79 n. 23; **1.400**: 79 n. 23

PHILOPONUS
in Physica **511,25**: 77 nn. 33-4; **541,22**: 81 n. 102; **550,10**: 84 n. 153; **550,18**: 91 n. 261; **550,17**: 84 n. 155; **565,21-566,7**: ix; **575,27-576,12**: 83 n. 134; **576,12-22**: 82 n. 128; **576,16-21**: 82 n. 129; **576,22-577,9**: 82 n. 129; **576,12-577,1**: viii; **577,3**: 91 n. 260

PLATO
Gorgias **465D5**: 92 n. 273
Laws **840D7**: 97 n. 348
Parmenides **138A3-7**: 77 n. 31
Phaedo **62A8**: 94 n. 311; **77E3**: 94 n. 311; **109A**: 92 n. 277
Philebus **18D6**: 89 n. 231
Republic **514A2-6**: 100 n. 388; **569C3-4**: 97 n. 348
Sophist **218D3**: 81 n. 96; **261A5**: 81 n. 96;
Symposium **174A3-4**: 80 n. 77
Theaetetus **168D3**: 76 n. 20
Timaeus **30A**: 100 n. 391; **39C5-D2**: 99 n. 383; **51A7-B1**: 78 n. 52, 79 n. 64; **52D4**: 100 n. 390; **61E3-4**: 94 n. 317; **62D**: 92 n. 277

PLOTINUS
3.7.9.62-3: 102 n. 418

SEXTUS EMPIRICUS
Math. **7.60**: 82 n. 120, 91 n. 262; **9.1**: 82 n. 120

Pyrrh. Hyp. **2.111**: 101 n. 404; **3.37**: 78 n. 45; **3.119**: 78 n. 45

SIMPLICIUS
in Physica **151,6-11**: 79 n. 63; **453,25-30**: 79 n. 63; **521,15-24**: 75 n. 4; **526,16-18**: 76 n. 11; **531,4-30**: 77 n. 31; **541,33-4**: 79 n. 62; **542,11**: 79 n. 64; **551,11-17**: 80 n. 90; **552,18-553,11**: 80 n. 74; **552,18-29**: 80 n. 92; **553,6-8**: 80 n. 74; **573,17-18**: 82 n. 123; **573,19**: 82 n. 124; **573,19-29**: 82 n. 128; **576,21-3**: 84 n. 159; **589,5-8**: 85 n. 164; **590,27-32**: ix; **592,11-593,6**: 86 n. 183; **592,22-593,6**: 86 n. 188; **592,25-7**: ix; **599,3-4**: 87 n. 200; **652,19-25**: 4 n. 7; **666,23-6**: 92 n. 277; **668,24-669,15**: 93 n. 289; **679,12-37**: 94 n. 318; **683,28-30**: 96 n. 333; **683,36-684,3**: 96 n. 335; **689,10**: 98 n. 355; **693,11-18**: 4 n. 7; **698,10-11**: 98 n. 361, 99 n. 370; **700,16-19**: 99 n. 383; **700,19-21**: 99 n. 380; **708,27-709,12**: 101 n. 407; **710,23-4**: 101 n. 414; **712,11-12**: 101 n. 418; **712,15-16**: 102 n. 421; **718,13-719,18**: viii; **723,33**: 104 n. 449; **732,26-733,1**: 105 n. 468; **733,16-18**: 105 n. 470; **744,6-8**: 106 n. 487; **751,4-5**: 107 n. 504; **753,10**: 108 n. 513; **754,7-17**: 108 n. 514; **754,9-13**: 108 n. 514; **754,26**: 108 n. 517; **754,30-1**: 108 n. 513; **755,2**: 109 n. 522; **759,18-760,3**: 109 n. 533; **762,11-24**: 110 n. 545; **765,32-766,19**: 110 n. 547; **766,6**: 110 n. 549; **766,8-9**: 110 n. 551; **766,13-19**: 110 n. 552; **768,13**: 111 n. 560

STOICORUM VETERUM FRAGMENTA
1.94 (Themist. *in Phys.* 123,20-2): 88 n. 214; **2.468** (Themist. *in Phys.* 104,13-18): 76 nn. 26-7; **2.552** (Simplic. *in Phys.* 671,4-12): 93 n. 296; **2.553** (Themist. *in Phys.* 130,13-17): 93 n. 296

THEMISTIUS
in Analytica Priora **1,16-17**: 94 n.
309; **1,16-2,4**: 84 n. 151; **1,2-12**:
4 n. 1; **29,20-3**: 76 n. 12
in de Anima **1,26**: 90 n. 246;
11,20-12,28: 79 n. 61; **18,30-7**:
86 n. 186; **44,31**: 92 n. 276;
49,1: 92 n. 276; **49,6.19.21**: 76
n. 16; **81,16**: 76 n. 16; **98,6**: 101
n. 402; **100,20-2**: 102 n. 428;
100,24-6: 110 n. 539; **110,34-6**:
106 n. 496

in de Caelo, **50,33-51**: 4 n. 7;
53,26-8: 88 n. 210
in Metaphysica Lambda (transl.
Brague) **para. 15**: 110 n. 537
in Physica **69,7-20**: viii;
81,30-82,18: 87 n. 190;
104,9-19 (*LS* 48F): 76 n. 26;
189,21ff.: 106 n. 490;
211,26-212,9: 100 n. 390;
234,12-235,29: 92 n. 285
Orations **1**: 106 n. 488; **23**
(89,20-90,5): 4 n. 1

Subject Index

References in brackets after proper names are to the page and line numbers of the Greek text; other proper names can be traced through the Index of Passages Cited. All other references are to the pages of the present work.

abstraction (mathematical), 76 n. 12
Alexander of Aphrodisias (104,20), on Stoic theory of total blending and bodily interprenetration, 77 n. 28; 88 n. 217; on Stoic cosmology, 93 n. 296
apergazesthai, sense of, 97 n. 350
Aquinas, Thomas, accepts Themistian interpretation of heavens being in place, 86 n. 187
Archytas (Pythagorean), 99 n. 380
Aristotle: exoteric writings, 98 n. 360; treatise 'On the Good', 78 n. 50; 79 n. 63; outermost sphere of universe being in place, ix, 86 nn. 187-8; prime matter, 78 n. 48; unmoved mover, 110 n. 537

Barbaro, Ermolao, translator of Themistius, 2 and n. 8
Boethus of Sidon (160,26; 163,6), on time, 109 nn. 533-4; 111 n. 556

change: of place, types of, 2; general relation to time, 55-6
Chrysippus of Soli (104,18; 113,11; 130,13), 76 n. 26; criticised for Stoic theory of cosmic stability, 43 and n. 296
Cleomedes, on Epicurean cosmology, 87 n. 191
clepsydras, used to disprove vacuum, 36 and n. 208, 46 and n. 320

Democritus of Abdera (123,7; 129,9; 142,31), on the void, 88 n. 212

ephelkesthai, 91 n. 273

Epicurus of Samos (106,6.8; 113,11; 123,17), 82 n. 121
Eudemus of Rhodes (119,26), on the heavens being in place, 80 n. 92; 86 n. 178

Galen of Pergamon (114,9.18; 144,24; 149,4), views on void and place criticised, viii, 82 nn. 128-9; 103 n. 439; on Aristotle on time, viii, 101 nn. 405, 407, 410; 103 n. 439; on time and change, 56; on time as self-defining, 60
growth, organic ('increase'), and the void, 37, 40; measurement by time, 71

Hesiod (103,28), concept of primal space, 76 n. 14
hexis, Stoic 'cohesive', 93 n. 296
hola, ta, translation of, 85 n. 173

impetus, theory of: not espoused by Themistius, 92 nn. 285, 289
in something, senses of, 22-3; something being in itself, 23-4

kermatizein, 96 n. 339
khoeus (liquid measure; 12 *kotulai*), 103 n. 434
khoinix (dry measure; 4 *kotulai*), 103 n. 434
khronos, translation of, 3
kinêsis/ kineisthai, translation of, 2-3, 109 n. 526
kotulê ('half pint'), 105 n. 471
kuathos (sixth of a *kotulê*), 96 n. 335

Leucippus (123,17), 88 n. 212

manuscripts (of Themistius):

Indexes

Florence (Biblioteca Medicea-Laurenziana 85,14), 101 n. 416; Modena (Biblioteca Estense *a*.M.9.13), 4 n. 9; Paris (Parisinus graecus 1886), 4 n. 9; Vatican City (Vaticanus graecus 1025), 75 n. 6; Wroclaw (Breslau) (Magd. 1442), 4 n. 9; Venice (Biblioteca Marciana graecus 205), 4 n. 9.

manos/puknos, translation of, 95 n. 328

matter, explains changes in density in preference to the void, 51; see place; prime, 78 n. 48

medimnos (corn measure), 105 n. 471

Melissus of Samos (124,3; 126,19), 89 n. 220

movement, impossible in the void, 42-5; natural and forced, 42; of projectiles, 42-3

mutual replacement (*antimetastasis*), primitive evidence of existence of place, 17; source of concept of place as extension, 27

now, the, senses of, 67; not a part of time, 53

ouranos, translation of, 78 n. 46, 87 n. 194, 88 n. 210

Paron (159,1; DK26), 108 n. 514

pêkhus ('24 finger-breadths'; 400 per stade), 98 n. 364; 105 n. 474; 111 n. 565

Philoponus, John, commentary on Aristotle's *Physics*, 2 and n. 5; defender of Galen on place, viii, 82 n. 129

phora/pheresthai, translation of, 2, 103 n. 448

place, recognised senses of, 25; implicitly two-dimensional, 76 n. 24, 84 n. 149; natural place, status of, 75 n. 9; not matter or form, 21-2, 26-7, 31; not three-dimensional extension, 27-31

Plato, and disorderly motion in the *Timaeus*, 100 nn. 390-1; as source of Themistian metaphors, 76 n. 20; 81 n. 96; 97 n. 348; unwritten doctrines and theory of matter as place, 78 n. 52; theory of void attributed to, 27 and n. 122

pote, translated as 'somewhen', 107 n. 499

Protagoras, 91 n. 262

selis (papyrus roll), 102 n. 428

Simonides of Ceos (158,27; fr. 19 Bergck), 108 n. 512

Simplicius, commentary on the *Physics*, 2; criticises Themistius, 96 n. 335; reproduces Themistius' problems on time, 110 nn. 547, 551-2

Socrates (109,5; 146,22(bis); 150,2.3); footwear, 80 n. 77; in Lyceum, 103 n. 443

Stoics, concept of common notions, 81 n. 97; infinite extra-cosmic void, 88 n. 214; use of *hormê*, 83 n. 140

Strato of Lampsacus, Themistius' knowledge of, 4 n. 7

Themistius, detection of fallacious reasoning, 75 n. 4; 82 n. 116; 103 n. 439; on putrefaction, 108 n. 518; on outermost sphere of universe being in place, ix, 86 nn. 187-8; paraphrastic method 1-2; problems formulated on time, 110 nn. 544-5, 547; pleonasm in, 80 n. 78; 92 n. 276

time, as conception of the mind, viii-ix, 70, 72; traditional cosmological conceptions of, 54-5; and the before and after, 58-9; and number and countability, 63-6; as cyclical, 72-3; measured by circular motion of heavens, 72; see also 'change' and 'now, the'

timelessness, 66-7

Vettori, Petro, emender of Themistius, 4 n. 10

void, *en masse* (separate and extra-cosmic), 88 n. 213; explanation of variations in density, 48-50; interstitial, 88 n. 211; undifferentiated, 27

wineskins, used by Anaxagoras

(123,9) to demonstrate resistance of air, 36; used by proponents of the void, 37 and n. 225

Xouthos (135,21.24; DK33), 96 nn. 331, 333, 335

Zeno of Citium (104,18; 123,21), 76 n. 26; 81 n. 94; and extra-cosmic void, 37 and n. 214

Zeno of Elea (105,12; 110,23), aporia on place, 77 n. 37, 80 n. 90, 81 n. 94, 89 n. 20